A Level
Mathematics
for Edexcel

Decision

D2

Brian Jefferson

OXFORD
UNIVERSITY PRESS

OXFORD
UNIVERSITY PRESS

Great Clarendon Street, Oxford OX2 6DP

Oxford University Press is a department of the University of Oxford.
It furthers the University's objective of excellence in research, scholarship,
and education by publishing worldwide in

Oxford   New York

Auckland   Cape Town   Dar es Salaam   Hong Kong   Karachi
Kuala Lumpur   Madrid   Melbourne   Mexico City   Nairobi
New Delhi   Shanghai   Taipei   Toronto

With offices in

Argentina   Austria   Brazil   Chile   Czech Republic   France   Greece
Guatemala   Hungary   Italy   Japan   South Korea   Poland   Portugal
Singapore   Switzerland   Thailand   Turkey   Ukraine   Vietnam

Oxford is a registered trade mark of Oxford University Press
in the UK and in certain other countries

British Library Cataloguing in Publication Data

Data available

ISBN 978-0-19-911785-7
10 9 8 7 6 5 4 3 2 1

Printed in Great Britain by Ashford Colour Press Ltd

Paper used in the production of this book is a natural,
recyclable product made from wood grown in sustainable forests.
The manufacturing process conforms to the environmental
regulations of the country of origin.

Acknowledgements

The Photograph on the cover is reproduced courtesy of iStockphoto

Series Managing Editor Anna Cox
The Publisher would like to thank the following for permission to reproduce photographs:
P20 Stephen Strathdee/Shutterstock; **P52** TebNad/Shutterstock; **P68** Michal Rosak/Shutterstock; **P88** Image
Source/Corbis; **P118** Ted Horowitz/Corbis; **P138** Libby Chapman/iStockphoto; **P156** Will
Iredale/Shutterstock.

The Publisher would also like to thank Ian Bettison, Naz Amlani and Charlie Bond
for their expert help in compiling this book.

About this book

Endorsed by Edexcel, this book is designed to help you achieve your best possible grade in Edexcel GCE Mathematics Decision 2 unit.

Each chapter starts with a list of objectives and an introduction. Chapters are structured into manageable sections, and there are certain features to look out for within each section:

Key points are highlighted in a blue panel.

Key words are highlighted in bold blue type.

Worked examples demonstrate the key skills and techniques you need to develop. These are shown in boxes and include prompts to guide you through the solutions.

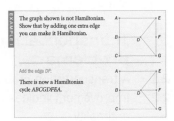

Derivations and additional information are shown in a panel.

Helpful hints are included as blue margin notes and sometimes as blue type within the main text.

| Misconceptions are shown in the right margin to help you avoid making common mistakes. | Investigational hints prompt you to explore a concept further. |

Each section includes an exercise with progressive questions, starting with basic practice and developing in difficulty. Some exercises also include 'stretch and challenge' questions marked with a stretch symbol .

At the end of each chapter there is a 'Review' section which includes exam style questions as well as past exam paper questions. There are also two 'Revision' sections per unit which contain questions spanning a range of topics to give you plenty of realistic exam practice.

The final page of each chapter gives a summary of the key points, fully cross-referenced to aid revision. Also, a 'Links' feature provides an engaging insight into how the mathematics you are studying is relevant to real life.

At the end of the book you will find full solutions, a key word glossary and an index.

Contents

1

Linear programming

This chapter will show you how to
- express an optimisation problem as a linear programming formulation (revision)
- solve two-variable problems graphically (revision)
- solve maximisation problems using the simplex algorithm.

Introduction

You met the main ideas of linear programming in the D1 unit, where you used graphical methods to solve problems with two decision variables. The application of linear programming in a real-world context often involves many variables. In such cases you use the algebraic methods described in this chapter.

Before you start

You should know how to:

1 Draw a straight line graph.

2 Draw the graph of a linear inequality, shading the unwanted region.

3 Solve linear simultaneous equations.

Check in:

1 Draw a graph showing the lines
$$y = 2x - 3 \quad \text{and} \quad 3x + 2y = 15$$
Verify that they intersect at $(3,3)$.

2 Draw a graph to show

  a $y \geqslant 0$

  b $2x + y \leqslant 15$

  c $y \leqslant 3x$

3 Solve these simultaneous equations.
$$x + y + z = 4$$
$$2x - 3y + 3z = 1$$
$$3x + 2y - z = -2$$

Linear programming is about using resources in the best way to maximise or minimise some quantity – typically to maximise profit or minimise cost.

To express the problem as a linear programming (LP) formulation:

- identify the quantities you can vary; these are the decision variables
- identify the limitations on the values of the decision variables; these are the constraints
- identify the quantity to be optimised; this is the objective function.

In this course you will not be required to deal with more than four variables.

EXAMPLE 1

A manufacturer makes three types of garden fertiliser – A, B and C. Each is a mixture of nitrate, phosphate and potash. The table shows the proportions of ingredients in the different types of fertiliser, the profit to be made on each and the amount of each ingredient available. You need to decide how much of each type to make in order to maximise profit.

	Nitrate (%)	Phosphate (%)	Potash (%)	Profit (£ per kg)
1 kg Type A	60	20	20	1.20
1 kg Type B	50	20	30	1.80
1 kg Type C	40	40	20	1.70
Availability (kg)	3000	1000	1200	

Write this problem as a linear programming formulation.

The decision variables are the amounts, x kg of Type A, y kg of Type B and z kg of Type C.

Take the constraints from the limitations on the amount of raw materials available:

Use $0.6x$ kg of nitrate for Type A, $0.5y$ kg for Type B and $0.4z$ kg for Type C.
There are 3000 kg of nitrate available, so $\qquad 0.6x + 0.5y + 0.4z \leqslant 3000$
giving $\qquad\qquad\qquad\qquad\qquad\qquad\qquad 6x + 5y + 4z \leqslant 30\,000$

Similarly, the availability of phosphate means that $\qquad x + y + 2z \leqslant 5000$
and the availability of potash means that $\qquad\qquad 2x + 3y + 2z \leqslant 12\,000$
There are also the non-negativity constraints, namely $\quad x \geqslant 0, y \geqslant 0, z \geqslant 0$

Check that you can see how these constraints are obtained.

The objective function in this case is the profit, £P, which you need to maximise:
The total profit is $\qquad P = 1.2x + 1.8y + 1.7z$

You can now state the complete linear programming formulation:

Maximise $\quad P = 1.2x + 1.8y + 1.7z$
Subject to $\qquad 6x + 5y + 4z \leqslant 30\,000$
$\qquad\qquad\qquad x + y + 2z \leqslant 5000$
$\qquad\qquad 2x + 3y + 2z \leqslant 12\,000$
$\qquad x \geqslant 0, y \geqslant 0, z \geqslant 0$

Exercise 1.1

For each of the following situations state the problem as a
linear programming formulation.

1   Moltobuono Meals Ltd make three different pasta sauces.
    Each is a blend of tomato paste and onion (of which they have
    unlimited quantities), with the addition of varying amounts of
    garlic paste, oregano and basil. This table gives details of the
    recipes (for 1 kg of sauce), together with the availability of
    ingredients and the profit on each type of sauce.

Sauce	Garlic (kg)	Oregano (kg)	Basil (kg)	Profit (£ per kg)
Assolato	0.03	0.02	0.01	0.60
Buona Salute	0.02	0.03	0.01	0.50
Contadino	0.03	0.04	0.02	0.90
Availability	50	60	40	

They wish to maximise their profit. Write the problem as a
linear programming formulation, simplifying the constraints
to have integer coefficients.

2   A liquid feed supplement for goats comprises concentrated
    glucose solution, vegetable oil and water. A litre of supplement
    must have no more than 600 ml of water. There must be no more
    than twice as much oil as glucose, and no more than three times
    as much glucose as oil. Glucose costs £1.40 per litre, oil costs 90p
    per litre and water costs 20p per litre. The manufacturer wants
    to minimise the cost of producing the supplement.
    Express the problem as a linear programming formulation.

3   A vegetable box scheme offers four different boxes, containing
    different proportions of baking potatoes, cabbages, swedes and
    butternut squashes. The table shows the proportions of these,
    the availability of the different vegetables and the profit.

Box	Potatoes	Cabbages	Swedes	Squashes	Profit (p per box)
A	6	2	1	1	50
B	4	1	1	2	30
C	8	2	2	2	80
D	5	1	2	1	60
Availability	800	240	300	320	

Express the problem of maximising profit as a linear programming formulation
(assume that all the boxes will be sold).

You can solve problems with two decision variables graphically.

Draw a graph showing the constraint inequalities (shading the unwanted regions), and identify the feasible region.

> The feasible region is the set of (x, y) values which satisfy all the constraints.
> It is the unshaded region on the graph.

You can illustrate the objective function on the graph by drawing an objective line.

> The objective line is a line joining all points (x, y) for which the objective function takes a specified value.

Your aim is to find the optimal position of the objective line. This occurs at a vertex of the feasible region (or along one boundary of the feasible region if the objective line is parallel to one of the constraints).
You can usually see from the graph which vertex is required. If not, you solve simultaneous equations to find the coordinates of the vertices and check by substitution which of them gives the optimal value for the objective function.

EXAMPLE 1

Maximise $\quad P = 3x + 2y$
subject to $\qquad x + y \leqslant 10$
$\qquad\qquad 2x + y \leqslant 16$
$\qquad\qquad x \geqslant 0, y \geqslant 0$

Draw the graph of the constraints, as shown.

Show an objective line. In this case the line

$$P = 3x + 2y = 15$$

has been drawn.

As the value of the objective function increases, the objective line moves to the right. Its most extreme position is when it passes through the intersection of $x + y = 10$ and $2x + y = 16$, which is the point $(6, 4)$.
This corresponds to $P = 26$.

So, P has a maximum value of 26, when $x = 6$ and $y = 4$.

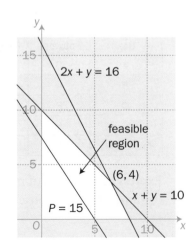

Exercise 1.2

1 For each of the following linear programming problems, draw a graph showing the feasible region, and one possible position of the objective line. Hence find the optimal value of the objective function and the corresponding values of x and y.

a Maximise $P = 3x + 4y$
 subject to $\quad 3x + 2y \leqslant 120$
 $\qquad\qquad x + 3y \leqslant 75$
 $\qquad\qquad x \geqslant 5, y \geqslant 10$

b Maximise $P = 4x + 3y$
 subject to $\quad x + y \leqslant 20$
 $\qquad\qquad 2x + y \leqslant 30$
 $\qquad\qquad x \geqslant 0, y \geqslant 0$

c Minimise $C = 4x + 5y$
 subject to $\quad 5x + 12y \geqslant 90$
 $\qquad\qquad 2y \geqslant x$
 $\qquad\qquad y \leqslant x$
 $\qquad\qquad x, y$ are integers

d Maximise $P = 5x + 4y$
 subject to $\quad x + y \leqslant 20$
 $\qquad\qquad 2x + 3y \leqslant 50$
 $\qquad\qquad 3x + 2y \leqslant 54$
 $\qquad\qquad x \geqslant 0, y \geqslant 0$

2 Solve graphically the minimisation problem described in Exercise 1.1 question **2**.

3 A manufacturer makes two types of juice. Frute-kup drink is 50% grape juice, 30% peach juice and 20% cranberry juice. Fresh-squeeze drink is 40% grape juice, 20% peach juice and 40% cranberry juice. Profit is 20p per litre for Frute-kup and 25p per litre for Fresh-squeeze. The manufacturer has 2100 litres of grape juice, 1200 litres of peach juice and 1500 litres of cranberry juice in stock. Use graphical methods to decide the most profitable quantities of the two drinks to make.

The optimal solution of a linear programming problem occurs at a vertex of the feasible region. The algebraic method tests whether a vertex gives the optimal solution and, if not, moves systematically to a better vertex until the problem is solved.

First rewrite the problem in terms of equations rather than inequalities. You do this by introducing slack variables.

Consider this linear programming formulation.

Maximise $P = 6x + 8y$

subject to $4x + 3y \leqslant 1500$

$x + 2y \leqslant 500$

$x \geqslant 0, y \geqslant 0$

Introduce variables $s = 1500 - (4x + 3y)$ and
$t = 500 - (x + 2y)$

s and t are the slack variables. They represent the 'spare capacity' in the two inequalities.

$s \geqslant 0$ and $t \geqslant 0$

The two main constraints become

$4x + 3y + s = 1500$

$x + 2y + t = 500$

Write the complete formulation:

Maximise $P = 6x + 8y$

subject to $4x + 3y + s = 1500$

$x + 2y + t = 500$

$x \geqslant 0, y \geqslant 0, s \geqslant 0, t \geqslant 0$

The line $4x + 3y = 1500$ is equivalent to $s = 0$.

The line $x + 2y = 500$ is equivalent to $t = 0$.

At each of the vertices O, A, B, and C of the feasible region, two of the variables x, y, s and t are zero.

The algebraic method effectively sets two of the variables to zero, finds the values of the other two variables and evaluates P.

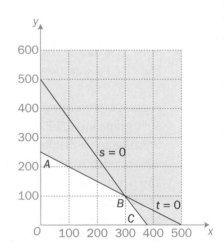

The following table shows the result of setting all possible pairs of x, y, s and t to zero.

For example, set $x = 0$, $t = 0$ (that is vertex A):
$$3y + s = 1500$$
$$2y = 500$$

Solve: $y = 250$, $s = 750$ and hence $P = 6 \times 0 + 8 \times 250 = 2000$
The complete table is:

Vertex	x	y	s	t	P
O	0	0	1500	500	0
A	0	250	750	0	2000
B	300	100	0	0	2600
C	375	0	0	125	2250
–	0	500	0	–500	–
–	500	0	–500	0	–

The optimal solution at B is $x = 300$, $y = 100$, $P = 2600$.

The last two rows do not correspond to vertices of the feasible region. This is shown by the negative values taken by the variables.

In each row of the table, the non-zero variables form the basis of a possible solution. They are the basic variables and the solution is a basic solution. Each of the first four rows is a basic feasible solution because the solution lies within the feasible region.

The zero variables are the non-basic variables.

You will use the simplex algorithm to search the basic solutions for the optimal solution.

The other two rows are basic non-feasible solutions.

Write each constraint as an equation, using a slack variable, in standard form, with all the variables on the left-hand side and a non-negative value on the right. Also write the objective function, P, in standard form, though the right-hand side may in this case be negative.

Enter the coefficients into a table called a simplex tableau.

Plural tableaux.

Write $\quad P - 6x - 8y - 0s - 0t = 0$
$$4x + 3y + s + 0t = 1500$$
$$x + 2y + 0s + t = 500$$
The simplex tableau is

Basic variable	x	y	s	t	Value
s	4	3	1	0	1500
t	1	2	0	1	500
P	–6	–8	0	0	0

This corresponds to the origin O in the feasible region, with $x = 0$, $y = 0$ giving $P = 0$.

Because $P - 6x - 8y = 0$, increasing either x or y will make P increase.

The basic variables always have a coefficient of 1 in their row, and zeros elsewhere in their column.

Read the table by looking at the first and last columns. $s = 1500$ and $t = 500$ (with x and y zero) give $P = 0$

A negative number in the objective row (P) tells you that the corresponding variable should be increased to reach the optimal solution.

> **Test for optimality** The table shows the optimal solution when the objective (bottom) row contains no negative coefficients.

The next step is to change the basis of the solution:

- Change one of the non-basic variables to a basic variable – it enters the basis. You are increasing its value.
- Change one of the basic variables to a non-basic variable – it leaves the basis. Its value becomes zero.

You must first choose which variables to change.

The variable entering the basis is the one whose entry in the objective row is the 'most negative'. The corresponding column is called the pivot column.

For each row divide the value by the entry in the pivot column. The results are called θ-values. The row giving the smallest θ-value is the pivot row. The basic variable in the pivot row is the one you choose to leave the basis.

The entry in the pivot row and pivot column is called the pivot.

y has the most negative entry in the objective row.
This column is the pivot column and y will enter the basis.

The θ-value for $s = \dfrac{1500}{3} = 500$, and for $t = \dfrac{500}{2} = 250$

It is smaller for t, so this is the pivot row, and t will leave the basis. The pivot is the 2 in the pivot column and pivot row.

Basic variable	x	y	s	t	Value	Row no.
s	4	3	1	0	1500	[1]
t	1	2	0	1	500	[2] ← pivot row
P	−6	−8	0	0	0	[3]

pivot column   pivot

1 Divide the pivot row by the pivot, 2. Change the basic variable from t to y.
2 In the y-column change the 3 to 0 by subtracting $3 \times y$-row.
3 In the y-column change the −8 to 0 by adding $8 \times y$-row.

*Any negative coefficient could indicate the pivot column.
Using the 'most negative' increases the chances of reaching the optimal solution quickly.*

This process identifies which of the basic variables will reach zero first. It avoids the possibility of reaching a non-feasible solution.

*For s-row, value = 1500 and entry in pivot column = 3.
For t-row, value = 500 and entry in pivot column = 2.*

You may find it helpful to put a 'Row no.' column so that you can record what has happened, as shown.

Remember that basic variables have a 1 in their row and zeros in the rest of their column. You need to make this true for y.

Take care to apply these row operations to the whole row.

Basic variable	x	y	s	t	Value	Row no.
s	$2\frac{1}{2}$	0	1	$-1\frac{1}{2}$	750	$[4]=[1]-3\ [5]$
y	$\frac{1}{2}$	1	0	$\frac{1}{2}$	250	$[5]=[2]\ 2$
P	−2	0	0	4	2000	$[6]=[3]+8\ [5]$

Always use fractions, not decimals, to avoid introducing rounding errors.

$s = 750$ and $y = 250$, giving $P = 2000$. Apply the optimality test.

The solution is optimal if there are no negative coefficients in the objective row. In this case there is −2 in the x-column, so you need to do another change of basis.

Row P is the objective row.

The x-column is the pivot column, so x will enter the basis.

The θ-value for $s = \dfrac{750}{2\frac{1}{2}} = 300$, and for $y = \dfrac{250}{\frac{1}{2}} = 500$

θ is smaller for s, so this is the pivot row, and s will leave the basis.

For s-row, value = 750 and entry in pivot column = $2\frac{1}{2}$.

For y-row, value = 250 and entry in pivot column = $\frac{1}{2}$.

Basic variable	x	y	s	t	Value	Row no.
s	$2\frac{1}{2}$	0	1	$-1\frac{1}{2}$	750	[4]
y	$\frac{1}{2}$	1	0	$\frac{1}{2}$	250	[5]
P	−2	0	0	4	2000	[6]

Divide the pivot row by $2\frac{1}{2}$.

Change the basic variable from s to x.

In the x-column change the $\frac{1}{2}$ to a 0 by subtracting $\frac{1}{2} \times x$-row.

In the x-column change the −2 to 0 by adding 2 x-row.

Basic variable	x	y	s	t	Value	Row no.
x	1	0	$\frac{2}{5}$	$-\frac{3}{5}$	300	$[7]=[4]\ 2\frac{1}{2}$
y	0	1	$-\frac{1}{5}$	$\frac{4}{5}$	100	$[8]=[5]-\frac{1}{2}\ [7]$
P	0	0	$\frac{4}{5}$	$2\frac{4}{5}$	2600	$[9]=[6]+2\ [7]$

You now have $x = 300$, $y = 100$, $s = 0$, $t = 0$, giving $P = 2600$. There are no negative coefficients in the objective row, so the solution is optimal.

D2

The simplex algorithm

Step 1   Write the constraints and the objective function as equations in standard form, using slack variables as needed.

Step 2   Transfer the data to a simplex tableau. At this stage the slack variables form the basis.

Step 3   Choose the column with the most negative coefficient in the objective row.
This is the pivot column.

Step 4   If the **positive** numbers in the pivot column are $p_1, p_2, \ldots$ and the corresponding numbers in the Value column are $v_1, v_2, \ldots$,
calculate $\theta_1 = \dfrac{v_1}{p_1}, \theta_2 = \dfrac{v_2}{p_2}, \ldots$.
The row giving the smallest θ-value is the pivot row. (If there is a tie, choose at random.) The number in the pivot column and pivot row is the pivot.

Step 5   Divide the pivot row by the pivot. Replace the basic variable for that row by the variable for the pivot column.

Step 6   Combine suitable multiples of the new pivot row with the other rows to give zeros in the pivot column.

Step 7   If there are no negative coefficients in the objective row, the solution is optimal.
Otherwise, go to Step 3.

The main advantage of the simplex algorithm compared to graphical methods is that you can use it when there are three or more decision variables.

EXAMPLE 1

Maximise $\quad P = x + 2y + z$

subject to $\quad x + 3y + 2z \leqslant 60$

$\qquad\qquad 2x + y \leqslant 40$

$\qquad\qquad x + 3z \leqslant 30$

Step 1 Write the problem in standard form with slack variables:

$$P - x - 2y - z = 0$$
$$x + 3y + 2z + s = 60$$
$$2x + y + t = 40$$
$$x + 3z + u = 30$$

Step 2 Transfer these data to a simplex tableau:

Basic variable	x	y	z	s	t	u	Value	Row no.
s	1	③	2	1	0	0	60	[1]
t	2	1	0	0	1	0	40	[2]
u	1	0	3	0	0	1	30	[3]
P	−1	−2	−1	0	0	0	0	[4]

Step 3 The y-column is the pivot column.

Step 4 The θ-values are: row [1]: $60 \div 3 = 20$

$\qquad\qquad\qquad\qquad\quad$ row [2]: $40 \div 1 = 40$

These give row [1] as the pivot row and the 3 as the pivot.

Row [3] cannot be the pivot row, as the pivot must be a positive number.

Step 5 Divide the pivot row by 3.

Replace s by y as the basic variable.

Step 6 Combine the other rows with suitable multiples of the new pivot row to get the required zeros:

Basic variable	x	y	z	s	t	u	Value	Row no.
y	$\frac{1}{3}$	1	$\frac{2}{3}$	$\frac{1}{3}$	0	0	20	$[5] = [1] \div 3$
t	$1\frac{2}{3}$	0	$-\frac{2}{3}$	$-\frac{1}{3}$	1	0	20	$[6] = [2] - [5]$
u	1	0	3	0	0	1	30	$[7] = [3]$
P	$-\frac{1}{3}$	0	$\frac{1}{3}$	$\frac{2}{3}$	0	0	40	$[8] = [4] + 2 \times [5]$

Example 1 is continued on the next page.

EXAMPLE 1 (CONT.)

Step 7 The solution is not optimal because there is a negative value in the bottom row.

Step 3 The x-column is the pivot column.

Step 4 The θ-values are:

row [5]: $20 \div \frac{1}{3} = 60$,

row [6]: $20 \div 1\frac{2}{3} = 12$,

row [7]: $30 \div 1 = 30$

These give row [6] as the pivot row and the $1\frac{2}{3}$ as the pivot.

Step 5 Divide the pivot row by $1\frac{2}{3}$.

Replace t by x as the basic variable.

Step 6 Combine the other rows with suitable multiples of the new pivot row to get the required zeros:

Basic variable	x	y	z	s	t	u	Value	Row no.
y	0	1	$\frac{4}{5}$	$\frac{2}{5}$	$-\frac{1}{5}$	0	16	$[9] = [5] - [10] \div 3$
x	1	0	$-\frac{2}{5}$	$-\frac{1}{5}$	$\frac{3}{5}$	0	12	$[10] = [6] \div 1\frac{2}{3}$
u	0	0	$3\frac{2}{5}$	$\frac{1}{5}$	$-\frac{3}{5}$	1	18	$[11] = [7] - [10]$
P	0	0	$\frac{1}{5}$	$\frac{3}{5}$	$\frac{1}{5}$	0	44	$[12] = [8] + [10] \div 3$

Step 7 There are no negative numbers in the bottom row. The solution is optimal.

You have $y = 16$, $x = 12$, $u = 18$, $z = 0$, $s = 0$, $t = 0$, giving $P = 44$.

You will be expected to be able to interpret each new tableau as you work through the algorithm.

EXAMPLE 2

A dog food manufacturer makes two types of food, each a mixture of meat and cereal. The ingredients and profit are shown in the table.

There are 2000 kg of meat and 1800 kg of cereal available.

The manufacturer seeks the most profitable production plan.

Type	Meat per kg	Cereal per kg	Profit (pence per kg)
A	0.6	0.4	20
B	0.5	0.5	15

a Express this as a linear programming formulation.

b Draw up a simplex tableau and perform one iteration of the simplex algorithm.

c Explain whether you have reached the optimal solution.

d Interpret the values of the variables at this stage.

a Let the production be x kg of type A and y kg of type B.

Maximise $P = 20x + 15y$

Subject to $0.6x + 0.5y \leqslant 2000$

$0.4x + 0.5y \leqslant 1800$

$x \geqslant 0, y \geqslant 0$

b $P - 20x - 15y = 0$

$0.6x + 0.5y + s = 2000$

$0.4x + 0.5y + t = 1800$

Basic variable	x	y	s	t	Value	Row no.
s	0.6	0.5	1	0	2000	[1]
t	0.4	0.5	0	1	1800	[2]
P	−20	−15	0	0	0	[3]

Basic variable	x	y	s	t	Value	Row no.
x	1	$\frac{5}{6}$	$1\frac{2}{3}$	0	$3333\frac{1}{3}$	$[4] = [1] \quad \frac{5}{3}$
t	0	$\frac{1}{6}$	$-\frac{2}{3}$	1	$466\frac{2}{3}$	$[5] = [2] - [4] \quad \frac{2}{5}$
P	0	$1\frac{2}{3}$	$33\frac{1}{3}$	0	$66666\frac{2}{3}$	$[6] = [3] + 20 \quad [4]$

Check that you can see how this second tableau was obtained.

c The solution is optimal because there are no negative numbers in the objective row.

d The values are $x = 3333\frac{1}{3}$, $t = 466\frac{2}{3}$, $y = 0$, $s = 0$, giving $P = 66666\frac{2}{3}$.

This means that the manufacturer should only make type A dog food in these circumstances. They can make $3333\frac{1}{3}$ kg of Type A. There will be $466\frac{2}{3}$ kg of cereal left at the end.

The simplex algorithm is designed to **maximise** the objective function.

You can use the algorithm to **minimise** the objective function. Minimising P is equivalent to maximising $(-P)$. You will **not** be asked to do this in the present syllabus.

Exercise 1.3

1 Use the simplex method to solve the following.

a Maximise $P = x + y$
subject to
$$x + 2y \leqslant 40$$
$$3x + 2y \leqslant 60$$
$$x \geqslant 0, y \geqslant 0$$

b Maximise $P = 2x + y$
subject to
$$4x + 3y \leqslant 170$$
$$5x + 2y \leqslant 160$$
$$x \geqslant 0, y \geqslant 0$$

c Maximise $P = 4x + 5y$
subject to
$$2x + 3y \leqslant 30$$
$$x + 3y \leqslant 24$$
$$4x + 3y \leqslant 48$$
$$x \geqslant 0, y \geqslant 0$$

d Maximise $P = x + 2y$
subject to
$$x + y \leqslant 6$$
$$2x + y \leqslant 9$$
$$3x + 2y \leqslant 15$$
$$x \geqslant 0, y \geqslant 0$$

e Maximise $P = 3x + 4y + 2z$
subject to
$$8x + 5y + 2z \leqslant 7$$
$$x + 2y + 3z \leqslant 4$$
$$x \geqslant 0, y \geqslant 0, z \geqslant 0$$

f Maximise $P = 7x + 6y + 4z$
subject to
$$2x + 4y - z \leqslant 7$$
$$5x + 6y + 2z \leqslant 16$$
$$7x + 7y + 4z \leqslant 25$$
$$x \geqslant 0, y \geqslant 0, z \geqslant 0$$

2 Pete Potts Horticultural Supplies sell two grades of grass seed, each a mixture of perennial ryegrass (PR) and creeping red fescue (CRF). Regular Lawn Mix is 70% PR and 30% CRF. Luxury Lawn Mix is equal proportions of each. Pete buys PR at £4 per kg and CRF at £5 per kg. He sells Regular mix at £6 per kg and Luxury mix at £7 per kg. He has 8000 kg of PR and 6000 kg of CRF in stock.

a Express the problem of maximising his profit as a linear programming formulation.

b Use the simplex algorithm to decide what quantities of Regular and Luxury mix he should make and state the profit he should receive.

3

Basic variable	x	y	z	s	t	Value
s	2	3	1	1	0	20
t	1	2	1	0	1	12
P	-3	-1	-2	0	0	0

a Write down the objective function and the constraints (in inequality form) corresponding to this simplex tableau.

b Perform one iteration of the simplex algorithm.
Explain how you know that the solution is not yet optimal.
State the values of the variables at this stage.

c Complete the solution of the problem, stating the final values of the variables.

4 You need to solve the following linear programming problem.

$$\text{Maximise} \quad P = x + 3y$$
$$\text{subject to} \quad x + y \leqslant 6$$
$$x + 4y \leqslant 12$$
$$x + 2y \leqslant 7$$
$$x \geqslant 0, y \geqslant 0$$

a Write the problem in terms of equations with slack variables.

b Enter the data into a simplex tableau. Perform one iteration of the simplex algorithm. Explain how you know that the optimal solution has not yet been reached.

c Perform a second iteration of the algorithm to obtain the optimal solution.

5 Solve the following linear programming problem using the simplex algorithm.

$$\text{Maximise} \quad P = 3x + 2y + z$$
$$\text{subject to} \quad x + y + 2z \leqslant 40$$
$$2x + 3y \leqslant 20$$
$$x + 2y + 2z \leqslant 30$$
$$x, y, z \geqslant 0$$

D2

6 A linear programming problem is written as a simplex tableau as follows:

Basic variable	x	y	s	t	Value
s	3	3	1	0	40
t	1	2	0	1	25
P	-5	-6	0	0	0

a Write down the original linear programming formulation, stating the objective function and showing the constraints as inequalities. Explain the meaning of the variables s and t in the tableau.

b Perform one iteration of the simplex algorithm. State the values of x, y, s, t and P at this stage, and explain how you know that the solution is not optimal.

c Perform a second iteration of the algorithm to obtain the optimal solution.

7 Brian's electric bicycle has two settings. On setting A he pedals with the motor providing assistance. On setting B the motor does all the work. Setting A gives a speed of $4\,\mathrm{m\,s}^{-1}$, using $6\,\mathrm{J}$ of battery energy per metre. Setting B gives a speed of $6\,\mathrm{m\,s}^{-1}$ and uses $9\,\mathrm{J}$ per metre. The battery can store $45\,000\,\mathrm{J}$. Brian wishes to travel as far as possible in 20 minutes.

a Show that if he travels x metres on setting A and y metres on setting B, the problem can be expressed in linear programming terms as follows:

Maximise $\quad D = x + y$

subject to $\quad 3x + 2y \leqslant 14\,400$

$\qquad\qquad\quad 2x + 3y \leqslant 15\,000$

$\qquad\qquad\quad x \geqslant 0, y \geqslant 0$

b Set up the corresponding simplex tableau. Hence find Brian's best distance and the settings he should use to achieve it.

8 Use the simplex algorithm to solve the problem described in Exercise 1.1, question **1**.

9 **a** Set up a simplex tableau for the problem described in Exercise 1.1, question **3**.

 b Use the simplex algorithm to obtain the optimal solution from your tableau.

 c Explain why the result obtained in part **b** does not provide a solution to the original problem. Use the result to arrive at a solution to the problem.

10 Consider the problem

Minimise $\quad C = x + y$

subject to $\quad x + 2y \geqslant 20$

$\qquad\qquad 3x + y \geqslant 30$

$\qquad\qquad x \geqslant 0, y \geqslant 0$

 a Write the two main constraints as equations involving non-negative slack variables s and t. In what way do these equations differ from those in a maximising problem?

 b The basic solution with x and y as the non-basic variables ($x = 0, y = 0$) is not feasible (s and t would be negative). Find the basic solution with y and s as the non-basic variables.

 c Express x, t and $(-C)$ in terms of y and s. Hence copy and complete the simplex tableau shown.

Basic variables	x	y	s	t	Value
x	1	2	–	0	20
t	0	–	-3	1	–
$(-C)$	0	–	–	0	–

 d Use the simplex algorithm to find the maximum value of $(-C)$ and hence the minimum value of C, together with the corresponding values of x and y.

D2

1 Bob's Bulbs Ltd sell three presentation boxes of tulip bulbs, A, B and C. Each contains packets of red, yellow and purple bulbs.

Box A contains 2 packets of red, 1 packet of yellow and 2 packets of purple bulbs.
Box B contains 1 packet of red, 3 packets of yellow and 1 packet of purple bulbs.
Box C contains 2 packets of red, 1 packet of yellow and 1 packet of purple bulbs.

There are 400 packets of red bulbs, 300 packets of yellow bulbs and 500 packets of purple bulbs available. They make £8 profit on a box of type A, £5 on B and £6 on C. They wish to maximise the profit. Let the numbers of box A, box B and box C made be x, y and z respectively.

a Formulate the above situation as a linear programming problem. State the objective function and list the constraints as equations using slack variables.

b Set up the corresponding simplex tableau and perform one iteration of the simplex algorithm. Explain how you know that the optimal solution has not yet been reached.

c **i** Perform one further iteration of the algorithm to obtain the optimal solution.

ii Interpret your current tableau, giving the value of each variable.

2 A company makes three models of hand-built bicycle. In a week they make x bicycles of type A, y of type B and z of type C. They make a weekly profit, in £hundreds, of P. In a certain week there are constraints in terms of resources and manpower on the number of bicycles they can produce and they want to decide how many of each type to make in order to maximise the profit. The simplex tableau corresponding to this problem is as shown.

Basic variable	x	y	z	s	t	Value
s	5	6	3	1	0	90
t	2	4	1	0	1	42
P	–4	–5	–2	0	0	0

a How much profit is made on a bicycle of type A?

b Write down the two constraints as inequalities.

c Use the simplex algorithm to solve this linear programming problem. Explain how you know when you have reached the optimal solution.

d Explain why the solution to part **c** is not practical.

e Find a practical solution which gives a profit of £7200. Verify that it is feasible.

3 A company makes three types of bed frame: *A*, *B* and *C*.
Each passes through three workshops for cutting, assembly and
finishing. The table shows the time, in worker-hours, that each
type needs in each workshop.

Type	Cutting	Assembly	Finishing	Profit per bed frame (£)
A	1	2	1	50
B	1	1	2	60
C	2	3	2	40
Worker-hours available	100	120	80	–

The company plans to make *x* of type *A*, *y* of type *B* and *z* of type *C*.

a Express the problem as a linear programming formulation,
  giving the constraints as equations with slack variables.

b Solve the problem by using the simplex algorithm.

c Interpret your solution.

4 Georgina is solving a linear programming problem using the
simplex algorithm. She obtains the following tableau.

Basic variable	x	y	z	s	t	u	Value
s	0	$1\frac{1}{2}$	0	1	0	-1	2
z	0	$2\frac{1}{3}$	1	0	1	$-\frac{1}{3}$	$\frac{1}{2}$
x	1	-1	0	0	-1	$\frac{1}{2}$	1
P	0	2	0	0	2	1	15

a Explain how she knows that she has reached the optimal solution.

b Write down the value of every variable.

c Write down the profit equation from the tableau.
  Explain why changing the value of any of the
  non-basic variables will decrease the value of *P*.

D2

Summary

Refer to

- Linear programming helps in planning how to use resources in the best way to maximise a profit or minimise a cost.   1.1
- You can solve two-variable problems graphically.
  - Draw a graph showing the constraints, and locate the feasible region – the set of points (x, y) satisfying all the constraints.
  - The optimal solution is at a vertex of the feasible region. Find this visually (draw an objective line) or test the coordinates of the vertices.   1.2
- You can find the solution to a maximisation problem algebraically using the simplex algorithm.
  - Express the constraints as equations with slack variables.
  - Enter the data in a simplex tableau. The basic variables (initially the slack variables) are non-zero at a particular vertex.
  - Test the optimality condition – the solution is optimal if there are no negative entries in the objective row.
  - If the solution is not optimal, change the basis, using row operations on the tableau.   1.3

Links

Linear programming can be used in the running of an airline. Applications within this setting can include:

- ensuring that every flight is covered
- ensuring that certain regulations are met (for example regarding the maximum permitted pilot flying time)
- minimising costs.

Linear programming techniques also allow the airline to build in as much flexibility as possible to minimise disruption from unexpected problems (such as aircraft breakdown or bad weather).

Airlines run on small profit margins, so any savings that can be made through good scheduling can make a crucial difference between profit and loss.

2

Transportation problems

This chapter will show you how to

- plan the distribution of a resource from several sources to several destinations
- decide whether a given distribution plan minimises the cost
- modify a given plan to reduce the cost
- express a transportation problem as a linear programming formulation.

Introduction

The usual linear programming problem involves the use of several resources, which you need to deploy in several combinations.

The problem you will meet in this chapter only involves one resource, which is available at several locations and is to be transported from these to several destinations. Although you can express the problem in conventional linear programming terms (see Section 2.6), this chapter shows a simpler, more efficient approach.

Suppose you want to transport chicken feed from suppliers
(sources) *A*, *B* and *C* to chicken farms (destinations) *X*, *Y* and *Z*.
You need to know three things:

- How much each source has in stock.
  This is called the supply, capacity or availability.
- How much each destination needs.
  This is called the demand or requirement.
- How much it costs to transport one unit of feed from each
  source to each destination.

The numbers of bags of feed available are:

A	B	C
60	90	70

The numbers of bags of feed needed are:

X	Y	Z
80	70	70

The cost of transport, in pounds per bag, is:

From \ To	X	Y	Z
A	3	5	4
B	6	2	5
C	4	7	3

You can show the cost information
as a bipartite graph.

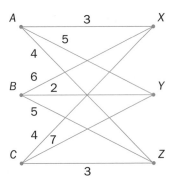

Put all the information into a **transportation tableau**:

From \ To	X	Y	Z	Supply (Availability)
A	3	5	4	60
B	6	2	5	90
C	4	7	3	70
Demand (Requirement)	80	70	70	220

Leave space in your table to
show your distribution plan.

The total demand and the total supply are both 220 bags, so
you have a balanced problem.

A transportation problem is balanced if the total supply equals the total demand.

You will meet unbalanced problems in Section 2.4.

Set up a possible distribution plan:

From \ To	X	Y	Z	Supply (Availability)
A	3 —	5 —	4 60	60
B	6 10	2 70	5 10	90
C	4 70	7 —	3 —	70
Demand (Requirement)	80	70	70	220

This shows, for example, that under this plan, B will supply 70 bags to Y, costing £2 each to transport.

The total cost of this distribution plan is:
$$4 \times 60 + 6 \times 10 + 2 \times 70 + 5 \times 10 + 4 \times 70 = £770$$

Look for ways to improve the plan. For example, you could save money by sending 10 bags to X from A rather than from B. This has a knock-on effect on other parts of the plan:

From \ To	X	Y	Z	Supply (Availability)
A	+10 3 —	5 —	−10 4 60	60
B	−10 6 10	2 70	+10 5 10	90
C	4 70	7 —	3 —	70
Demand (Requirement)	80	70	70	220

If A sends 10 bags to X, then only 50 bags can go to Z, so that A's capacity stays at 60. This in turn alters what B sends to Z. This loop of adjustments is called the **stepping-stones method**. This is explored more fully in Section 2.3.

The new plan looks like this:

From \ To	X	Y	Z	Supply (Availability)
A	3 10	5 —	4 50	60
B	6 —	2 70	5 20	90
C	4 70	7 —	3 —	70
Demand (Requirement)	80	70	70	220

The total cost of this distribution plan is:
$$3 \times 10 + 4 \times 50 + 2 \times 70 + 5 \times 20 + 4 \times 70 = £750$$

You can make a second improvement, as shown:

From \ To	X	Y	Z	Supply (Availability)
A	+50 ⟨3⟩ 10	⟨5⟩ –	−50 ⟨4⟩ 50	60
B	⟨6⟩ –	⟨2⟩ 70	⟨5⟩ 20	90
C	−50 ⟨4⟩ 70	⟨7⟩ –	+50 ⟨3⟩ –	70
Demand (Requirement)	80	70	70	220

The distribution plan now looks like this:

From \ To	X	Y	Z	Supply (Availability)
A	⟨3⟩ 60	⟨5⟩ –	⟨4⟩ –	60
B	⟨6⟩ –	⟨2⟩ 70	⟨5⟩ 20	90
C	⟨4⟩ 20	⟨7⟩ –	⟨3⟩ 50	70
Demand (Requirement)	80	70	70	220

The total cost of this distribution plan is:
$$3 \times 60 + 2 \times 70 + 5 \times 20 + 4 \times 20 + 3 \times 50 = £650$$

This is the cheapest plan for this situation.

You will meet the method for deciding whether the best solution has been reached in Section 2.2.

You are expected to be able to use the north-west corner method to set up an initial distribution plan.

The north-west corner method

Start with the top left-hand cell of the tableau.

Step 1   Allocate as many units as possible/necessary to the current cell. Stop if the table is complete.

Step 2   If supply in the current row is exhausted, move to the row below.

Step 3   If demand in the current column is met, move to the column on right.

Step 4   Go to Step 1.

Applying the north-west corner method to the chicken feed problem gives:

Step 1   Allocate 60 from *A* to *X*.
Step 2   Supply in first row exhausted so move to cell below.
Step 3   Demand not met.

To / From	X	Y	Z	Supply (Availability)
A	3 **60**	5 –	4 –	60
B	6	2	5	90
C	4	7	3	70
Demand (Requirement)	80	70	70	220

Step 1   Allocate 20 from *B* to *X*.
Step 2   Supply not exhausted.
Step 3   Demand in first column met so move to cell on right.
Step 1   Allocate 70 from *B* to *Y*.

To / From	X	Y	Z	Supply (Availability)
A	3 **60**	5 –	4 –	60
B	6 **20**	2 **70**	5	90
C	4	7	3	70
Demand (Requirement)	80	70	70	220

Step 2   Supply in second row exhausted so move to cell below.
Step 3   Demand in second column met so move to cell on right.
Step 1   Allocate 70 from *C* to *Z*. Table complete, so stop.

To / From	X	Y	Z	Supply (Availability)
A	3 **60**	5 –	4 –	60
B	6 **20**	2 **70**	5 –	90
C	4 –	7 –	3 **70**	70
Demand (Requirement)	80	70	70	220

The cost of this allocation is £650, which you saw earlier is the minimum cost of allocating these resources.

The north-west corner method will not usually produce an optimal solution.

Exercise 2.1

1 For each of the following transportation tableaux, use the north-west corner method to obtain an initial allocation.

a

To From	X	Y	Z	Supply
A				200
B				140
C				230
Demand	210	240	120	570

b

To From	X	Y	Z	Supply
A				100
B				130
C				90
Demand	80	150	90	320

c

To From	P	Q	R	S	Supply
A					960
B					750
C					820
Demand	850	590	670	420	2530

d

To From	X	Y	Z	Supply
A				60
B				80
C				110
D				120
Demand	150	90	130	370

2 In a certain area there are three builders' merchants, A, B and C, supplying bags of cement, and three major building projects, X, Y and Z.

The numbers of bags the merchants have in stock are as follows:

A	B	C
120	95	145

The requirements of the building projects are as follows:

X	Y	Z
85	150	125

The costs, in £ per bag, of transporting the cement are as follows:

To From	X	Y	Z
A	2	3	2
B	4	6	3
C	2	4	4

a Explain why this is a balanced problem.

b Draw up a transportation tableau and find an initial allocation of resources using the north-west corner method.

c Calculate the cost of your initial allocation.

You are going to test whether a given allocation is optimal.

> In the transportation tableau
> - Cell (i, j) refers to the cell in the ith row and jth column
> - C_{ij} refers to the unit cost in cell (i, j).

Assign values, called shadow costs, to each row and column.

> - R_i is the shadow cost for the ith row.
> - K_j is the shadow cost for the jth column.

For a table with m rows and n columns, these shadow costs can only be found if there are $(m + n - 1)$ occupied cells. If this is not so, the solution is said to be **degenerate**. You will meet degenerate situations in Section 2.5.

The shadow costs are chosen so that for each occupied cell

$$R_i + K_j = C_{ij}$$

This is the initial allocation for the chicken feed problem.

From \ To	X	Y	Z	Supply (Availability)
A	[3] —	[5] —	[4] 60	60
B	[6] 10	[2] 70	[5] 10	90
C	[4] 70	[7] —	[3] —	70
Demand (Requirement)	80	70	70	220

The unit costs of the occupied cells are
$C_{13} = 4$, $C_{21} = 6$, $C_{22} = 2$, $C_{23} = 5$ and $C_{31} = 4$

So you need

$$R_1 + K_3 = 4 \quad [1]$$
$$R_2 + K_1 = 6 \quad [2]$$
$$R_2 + K_2 = 2 \quad [3]$$
$$R_2 + K_3 = 5 \quad [4]$$
$$R_3 + K_1 = 4 \quad [5]$$

There are five equations and six unknowns, so if you choose a value for one of the unknowns you can calculate the corresponding values of the others.

This explains why degeneracy is a problem. Fewer equations would prevent you from calculating these shadow costs.

It is usual to start by setting $R_1 = 0$.

From [1] you get $K_3 = 4$
From [4] you get $R_2 = 1$
From [2] you get $K_1 = 5$
From [3] you get $K_2 = 1$
From [5] you get $R_3 = -1$

You can now use your shadow costs to calculate a predicted unit cost for each unoccupied cell (i, j).

Predicted unit cost $= R_i + K_j$

The difference between this predicted cost and the actual unit cost C_{ij} for that cell tells you how much the total cost would change if one unit were allocated to cell (i, j). This is called the improvement index.

The improvement index for cell (i, j) is
$$I_{ij} = C_{ij} - R_i - K_j$$

By definition the improvement index for an occupied cell is zero.

If $I_{ij} < 0$ then allocating units to cell (i, j) would reduce the total cost. It follows that

An allocation is optimal if and only if $I_{ij} \geqslant 0$ for all unoccupied cells.

Returning to the chicken feed problem, you have

$I_{11} = 3 - 0 - 5 = -2$
$I_{12} = 5 - 0 - 1 = 4$
$I_{32} = 7 - (-1) - 1 = 7$
$I_{33} = 3 - (-1) - 4 = 0$

The allocation is not optimal. Changing the solution by allocating units to cell $(1, 1)$ would reduce the total cost by £2 for each unit allocated.

This is in fact what you did on page 23. Ten units were reallocated to cell $(1, 1)$ and the total cost reduced from £770 to £750.

D2

In practice you may find it convenient to use a table when calculating shadow costs and improvement indices.

EXAMPLE 1

Calculate improvement indices for this transportation tableau, and state with reasons whether the allocation is optimal.

To From	X	Y	Z	Supply
A	5 70	9 –	6 –	70
B	3 20	6 20	5 –	40
C	2 –	4 10	2 50	60
Demand	90	30	50	170

Draw up a table to show shadow costs and improvement indices:

	$K_1 =$	$K_2 =$	$K_3 =$
$R_1 =$	5 X	9	6
$R_2 =$	3 X	6 X	5
$R_3 =$	2	4 X	2 X

You can identify the occupied cells with Xs as shown.

Set $R_1 = 0$. It follows that $K_1 = 5$ (because $C_{11} = 5$)
This in turn gives $R_2 = -2$ (because $C_{21} = 3$)
Continuing in this way, you find all the shadow costs:

	$K_1 = 5$	$K_2 = 8$	$K_3 = 6$
$R_1 = 0$	5 X	9	6
$R_2 = -2$	3 X	6 X	5
$R_3 = -4$	2	4 X	2 X

Check that you can see where these values come from.

Calculate the improvement indices:

$$I_{ij} = C_{ij} - R_i - K_j$$

So, for example, $I_{12} = 9 - 0 - 8 = 1$
and $I_{23} = 5 - (-2) - 6 = 1$

EXAMPLE 1 (CONT.)

Record the results in the table. You may wish to circle them, to remind yourself that these are improvement indices, not allocations.

	$K_1 = 5$	$K_2 = 8$	$K_3 = 6$
$R_1 = 0$	5 X	9 ①	6 ⓪
$R_2 = -2$	3 X	6 X	5 ①
$R_3 = -4$	2 ①	4 X	2 X

There are no negative improvement indices, so the allocation is optimal.

There is no universally agreed notation for showing this working. The important thing is that your answer should show the shadow costs and the improvement indices you obtain.

Check that you can see where these values come from.

Exercise 2.2

1 By calculating the improvement index for each unoccupied cell, test whether the following allocations are optimal.

a

From \ To	X	Y	Z	Supply
A	6 40	4 10	5 –	50
B	7 –	8 30	8 –	30
C	7 –	5 20	3 50	70
Demand	40	60	50	150

b

From \ To	X	Y	Z	Supply
A	3 160	3 –	4 –	160
B	7 –	4 180	6 70	250
C	4 50	5 –	5 150	200
Demand	210	180	220	610

D2

c

From \ To	X	Y	Z	Supply
A	[6] 80	[9] –	[6] –	80
B	[3] –	[3] 70	[2] –	70
C	[2] 60	[3] –	[2] 50	110
D	[3] –	[5] 80	[3] 50	130
Demand	140	150	100	3900

d

From \ To	P	Q	R	S	Supply
A	[5] –	[7] –	[4] –	[8] 120	120
B	[7] –	[9] –	[4] 80	[9] 10	90
C	[4] –	[7] 65	[3] 35	[9] –	100
D	[2] 80	[5] 60	[2] –	[7] –	140
Demand	80	125	115	130	450

2 Use the north-west corner method to find an initial allocation for this transportation tableau. Show that the allocation you obtain is optimal.

From \ To	X	Y	Z	Supply
A	[4]	[3]	[4]	90
B	[7]	[5]	[6]	130
C	[6]	[4]	[5]	80
Demand	70	80	150	300

3 A company stores sand at three depots, *A*, *B* and *C*. It requires
sand at three sites, *X*, *Y* and *Z*. The lengths, in km, of round
trips between depots and sites are given by the weights in this
bipartite graph, which also shows the number of lorry loads
available at the depots and required by the sites.

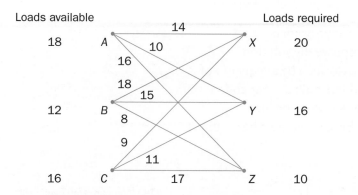

a Show this data in a transportation tableau.

b Use the north-west corner method to find an initial allocation.

c Calculate the total distance travelled by the lorries for the
allocation found in part **b**.

d Calculate improvement indices and hence state whether this
allocation is optimal.

4

To / From	X	Y	Z	Supply
A	7	7	1	100
B	2	6	5	160
C	3	4	6	120
Demand	80	190	110	380

a Use the north-west corner method to find an initial allocation
for this transportation tableau, and show that the solution
obtained is not optimal.

b By starting in the bottom left-hand cell (the south-west corner method),
obtain a second allocation. Show that this, too, is not optimal.

c A third possible approach is the least-cost method, in which
the cells are used in increasing order of their unit costs.
Use this method to obtain another allocation, and show that
this allocation is optimal.

D2

Changing the allocation of one cell has a knock-on effect, requiring a sequence of adjustments to other cells in a loop. This is the stepping-stones method and is defined as follows:

Refer to Section 2.1.

- First identify a cell with a negative improvement index. Allocating units to this cell will improve the solution.
- Then identify a loop (closed path), formed by a sequence of steps from cell to cell, starting and finishing at your chosen cell.
  - Each step must be at right angles to the previous step.
  - All the cells, apart from your chosen cell, must be occupied.

Here are three examples:

The loop can 'step over' cells, and need not be a rectangle.

You wish to allocate as many units as possible to cell $(3, 1)$ in this table.

Identify a loop, as shown.

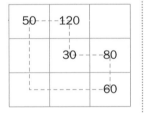

Suppose you allocate x units to cell $(3, 1)$.
To preserve column and row totals you move round the loop, alternately decreasing and increasing the allocations by x units, as shown.

If cell $(3, 1)$ increases by x, then cell $(1, 1)$ must decrease by x so that column 1 total is unchanged. This in turn means that cell $(1, 2)$ must increase by x so that row 1 total is unchanged. And so on.

You cannot allow any cell allocation to become negative, so the largest value of x is 30. This would completely empty cell $(2, 2)$.

Cell $(3, 1)$ has entered the allocation scheme. It is called the **entering cell** or **entering route**. Cell $(2, 2)$ has left the allocation scheme. It is called the **exiting cell** or **exiting route**. In the exam you should state the entering and exiting cells used in your answer.

Using $x = 30$, the revised allocation is as shown.

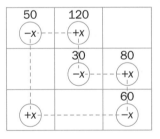

From \ To	X	Y	Z	Supply
A	[4] 200	[2] –	[5] –	200
B	[5] 20	[2] 220	[4] –	240
C	[6] –	[3] 40	[6] 120	160
Demand	220	260	120	600

a Starting with the initial allocation shown, find an optimal allocation for this transportation tableau.

b Show that there is a second optimal allocation.

The total cost of this initial allocation is £2180.

a First find the shadow costs and hence the improvement indices:

	$K_1 = 4$	$K_2 = 1$	$K_3 = 4$
$R_1 = 0$	[4] X	[2] ①	[5] ①
$R_2 = 1$	[5] X	[2] X	[4] (−1)
$R_3 = 2$	[6] ⓪	[3] X	[6] X

The improvement indices are shown circled in the table.

Allocating units to cell $(2,3)$ will give an improved allocation.

Each unit moved to cell $(2,3)$ will reduce the total cost by £1.

Identify a loop, as shown:

To avoid negative values, the maximum value of x is 120.

The entering cell is $(2,3)$.
The exiting cell is $(3,3)$.

200		
20	220 (−x)	(+x)
	40 (+x)	120 (−x)

If there are two or more cells with negative improvement indices you would usually choose the cell with the 'most negative' index as your entering cell, as this is likely to produce the greatest cost benefit and lead you more quickly to the optimal solution. Choose at random if there is a tie.

Using $x = 120$, the revised allocation is as shown:

200		
20	100	120
	160	

Recalculate the shadow costs and improvement indices for the new allocation:

	$K_1 = 4$	$K_2 = 1$	$K_3 = 3$
$R_1 = 0$	[4] X	[2] ①	[5] ②
$R_2 = 1$	[5] X	[2] X	[4] X
$R_3 = 2$	[6] ⓪	[3] X	[6] ①

Example 1 is continued on the next page.

D2

35

EXAMPLE 1 (CONT.)

There is now no negative improvement index, so the allocation is optimal.

The final table is as shown:

To From	X	Y	Z	Supply
A	4 200	2 –	5 –	200
B	5 20	2 100	4 120	240
C	6 –	3 160	6 –	160
Demand	220	260	120	600

If the allocation is not yet optimal, repeat the process to improve it.

The total cost of this optimal allocation is £2060 (that is, it has improved by £120, as expected).

b The improvement index for cell $(3, 1)$ is 0. This shows that units can be allocated to this cell without changing the total cost.

The entering cell is $(3, 1)$, which receives the maximum 20 units, as shown.
The exiting cell is $(2, 1)$

200		
20 −20	100 +20	120
+20	160 −20	

There are in fact a number of optimal allocations. You might, for example, have only allocated 15 to cell $(3, 1)$, leaving cell $(2, 1)$ as 5, cell $(2, 2)$ as 115 and cell $(3, 2)$ as 145. It would, however, be unusual to use more than the minimum necessary number of cells in an allocation.

This is the corresponding transportation tableau:

To From	X	Y	Z	Supply
A	4 200	2 –	5 –	200
B	5 –	2 120	4 120	240
C	6 20	3 140	6 –	160
Demand	220	260	120	600

The total cost is still £2060, so this is also an optimal allocation.

Exercise 2.3

1 Copy the following tables and alter them by allocating as many units as possible to the cell marked with an asterisk.
State the entering and exiting cells in each case.

a

100		*
60	80	50
	140	

b

170	150	
	90	*
	100	230

c

700			*
400	260		
	500		
	220	150	300

2 a Use the north-west corner method to find an initial allocation for this transportation tableau.
Calculate the total cost of this allocation.

b Improve on your initial allocation to find an optimal allocation. Calculate the cost of this optimal solution, and explain how you know that it is optimal.

From \ To	X	Y	Z	Supply
A	8	6	8	230
B	6	3	4	350
C	3	4	3	200
Demand	300	210	270	780

3

From \ To	P	Q	R	S	Supply
A	4	2	7	4	80
B	5	2	5	3	110
C	2	4	5	2	60
Demand	70	40	40	100	250

a Use the north-west corner method to find an initial allocation for this transportation tableau.

b Improve your initial allocation to find two optimal solutions to the problem, each using six cells of the table.
State the cost of the optimal solution.

A transportation problem is unbalanced if the number of units required (the demand) is not equal to the number of units available (the supply).

You can turn an unbalanced table into a balanced table by introducing a dummy source or destination. The unit costs in the dummy row or column will be zero because the units will not actually be delivered, so no costs will be incurred.

> If demand is greater than supply, some destinations will not receive all the units they require. If supply is greater than demand, some sources will have units left over.

EXAMPLE 1

From \ To	X	Y	Z	Supply
A	4	3	6	80
B	2	4	2	100
Demand	60	90	70	

a Rewrite the tableau shown to give a balanced problem and find an initial allocation.

b Find the optimal allocation and interpret the result.

a There are 180 units available and 220 units are required. Introduce a dummy source to 'supply' the missing 40 units.

The north-west corner method then gives the initial allocation shown:

From \ To	X	Y	Z	Supply
A	4 60	3 20	6 –	80
B	2 –	4 70	2 30	100
Dummy	0 –	0 –	0 40	40
Demand	60	90	70	220

b Calculate the improvement indices, as shown.

	$K_1 = 4$	$K_2 = 3$	$K_3 = 1$
$R_1 = 0$	4 X	3 X	6 (5)
$R_2 = 1$	2 (−3)	4 X	2 X
$R_3 = −1$	0 (−3)	0 (−2)	0 X

> Choose cell $(2, 1)$ as the entering cell (it ties with $(3, 1)$ for the 'most negative' index).

EXAMPLE 1 (CONT.)

A maximum of 60 units can be allocated to cell $(2,1)$, using the loop shown.

60 ⊖−60	20 ⊕+60	
⊕+60	70 ⊖−60	30
		40

Cell $(1,1)$ is the exiting cell.

This gives the revised tableau shown.

From \ To	X	Y	Z
A	[4] –	[3] 80	[6] –
B	[2] 60	[4] 10	[2] 30
Dummy	[0] –	[0] –	[0] 40

Recalculate the improvement indices for the revised tableau, as shown.
The solution is not yet optimal.

	$K_1 = 1$	$K_2 = 3$	$K_3 = 1$
$R_1 = 0$	[4] ③	[3] X	[6] ⑤
$R_2 = 1$	[2] X	[4] X	[2] X
$R_3 = -1$	[0] ⓪	[0] ⓐ−2	[0] X

Cell $(3,2)$ is the entering cell.

A maximum of 10 units can be allocated to cell $(3,2)$, using the loop shown.

	80	
60	10 ⊖−10	30 ⊕+10
	⊕+10	40 ⊖−10

Cell $(2,2)$ is the exiting cell.

This gives the revised tableau shown:

From \ To	X	Y	Z
A	[4] –	[3] 80	[6] –
B	[2] 60	[4] –	[2] 40
Dummy	[0] –	[0] 10	[0] 30

Example 1 is continued on the next page.

EXAMPLE 1 (CONT.)

Recalculate the improvement indices once again:

	$K_1 = 3$	$K_2 = 3$	$K_3 = 3$
$R_1 = 0$	4 ①	3 X	6 ③
$R_2 = -1$	2 X	4 ②	2 X
$R_3 = -3$	0 ⓪	0 X	0 X

None of the indices is negative, so an optimal solution has been reached.

The values in the dummy row show the shortfall in supply. Destination Y will be 10 units short of the 90 units required, and destination Z will be 30 units short of the 70 units required.

In Example 1 the demand was greater than the supply, so a dummy source was used to balance the problem.

If the supply were greater than the demand, you would need to introduce an extra column for a dummy destination to 'receive' the extra units.

Exercise 2.4

1 This table shows the units available from two suppliers A and B, and the units needed by two retail outlets X and Y. Unit costs are in £.

  a Rewrite the table using a dummy to create a balanced problem.

  b Show that the minimum cost for which all the required units can be supplied is £1330.

  c State how many surplus units each supplier will have left if the optimal allocation is used.

To \ From	X	Y	Supply
A	4	7	200
B	3	5	100
Demand	120	150	

2 The transportation tableau shows the number of new games consoles ordered by three shops X, Y and Z, and the number of consoles in stock at three wholesalers A, B and C. The unit costs are in £.

  a Introduce a dummy to create a balanced problem.

  b Find the minimum cost for which the allocation can be made.

  c There are a number of ways in which a minimum cost allocation can be made.

    i Which shop is bound to have a shortfall in the number of consoles they receive? What is the least shortfall they could have?
    ii Which shop is bound to receive all the consoles they ordered?

To \ From	X	Y	Z	Supply
A	4	3	4	260
B	5	2	6	320
C	7	3	5	190
Demand	300	360	240	

3 The table shows the cost, in £, of transporting a pallet of bricks from each of three depots A, B and C, to each of two building sites X and Y. On a certain day A, B and C have respectively 90, 70 and 40 pallets of bricks in stock. Site X needs 80 pallets and site Y needs 60 pallets.

From \ To	X	Y
A	50	70
B	30	10
C	60	40

a Find the minimum cost of meeting the sites' requirements.

b How many pallets will be left in stock at each of the depots?

4 Two fruit farms, X and Y, need respectively 75 and 94 temporary workers at the start of the fruit picking season. The agencies they use, A and B, have respectively 84 and 56 workers available. The costs (in £ per worker) of transport from agency to farm are shown in the table.

From \ To	X	Y
A	10	7
B	4	8

a Draw up a transportation tableau to show the corresponding balanced problem. Use the north-west corner method to obtain an initial allocation.

b Calculate shadow costs and improvement indices for the initial allocation. Explain how you know the solution is not optimal.

c Stating the entering and exiting cell at each stage, make successive improvements to the allocation until the optimal allocation is reached.

d By how many workers will each farm be short of their required workforce?

5 The table shows the number of cases of lemonade in stock at three wholesalers, A, B and C, and the number of cases ordered by three shops, X, Y and Z. The costs shown are the costs, in £ per ten cases, of transporting the lemonade.

From \ To	X	Y	Z	Supply
A	3	3	3	180
B	4	2	6	100
C	6	5	4	170
Demand	130	180	200	

a Use the north-west corner method to find an initial solution for the corresponding balanced problem.

b Improve your initial solution to find an optimal allocation. Calculate the cost of this solution.

c Show that it is possible to allocate the units at minimum cost in such a way that the shortfalls in the orders from the three shops are equal.

D2

EXAMPLE 1

2.5 Degeneracy

For a table with m rows and n columns there are $(m + n)$ shadow cost values to find. By choosing one of these (usually $R_1 = 0$), you can calculate the remaining $(m + n - 1)$ values. You need $(m + n - 1)$ independent equations, each corresponding to an occupied cell.

Sometimes you have a table with fewer than $(m + n - 1)$ occupied cells. This may happen with the initial allocation, or as a result of using the stepping-stones method. Either way, the allocation is said to be degenerate.

> A solution to a transportation problem with m rows and n columns is degenerate if the number of occupied cells is less than $(m + n - 1)$.

The remedy is to artificially create one or more occupied cells so that the number of occupied cells is $(m + n - 1)$. You do this by assigning zero units to your chosen cells and treating them as being occupied.

You must choose the cell or cells so that the R and K values can all be found.

Use the north-west corner method to find an initial solution to this transportation problem. Hence find the optimal allocation.

From \ To	X	Y	Z	Supply
A	3	5	4	60
B	6	2	8	90
C	1	7	3	70
Demand	80	70	70	220

The north-west corner method allocates the units as shown.

From \ To	X	Y	Z	Supply
A	3 60	5 –	4 –	60
B	6 20	2 70	8 –	90
C	1 –	7 –	3 70	70
Demand	80	70	70	220

Here $m = n = 3$, so $m + n - 1 = 5$. There are only four occupied cells, so the solution is degenerate.

Creating an occupied cell in either row 3 or column 3 will remedy the situation.
Create an occupied cell at $(1, 3)$.

This happens because when the 70 is allocated to cell $(2, 2)$ it completes the totals for both row 2 and column 2.

You can not use cell $(1, 2)$ here, as the 70 in cell $(3, 3)$ would still be isolated, with two unknown shadow costs and no means of finding either.

EXAMPLE 1 (CONT.)

Calculate the shadow costs and improvement indices:

	$K_1 = 3$	$K_2 = -1$	$K_3 = 4$
$R_1 = 0$	3 — X	5 — ⑥	4 — X
$R_2 = 3$	6 — X	2 — X	8 — ①
$R_3 = -1$	1 — ⑴₋₁	7 — ⑨	3 — X

Improve the solution using the stepping-stones method:

60 ⊖−60		0 ⊕+60
20	70	
⊕+60		70 ⊖−60

Here is the revised allocation:

From \ To	X	Y	Z
A	3 —	5 — –	4 — 60
B	6 — 20	2 — 70	8 — –
C	1 — 60	7 — –	3 — 10

Recalculate the shadow costs and improvement indices:

	$K_1 = 2$	$K_2 = -2$	$K_3 = 4$
$R_1 = 0$	3 — ①	5 — ⑦	4 — X
$R_2 = 4$	6 — X	2 — X	8 — ⓪
$R_3 = -1$	1 — X	7 — ⑩	3 — X

There are now no negative improvement indices, so the solution is optimal.

Try repeating this example choosing another cell as the extra occupied cell.

D2

A degenerate situation arises with the stepping-stones method when there are two or more exiting cells in the loop, as in Example 2.

EXAMPLE 2

From \ To	X	Y	Z	Supply
A	[2] 140	[5] –	[1] –	140
B	[4] 50	[6] 80	[5] –	130
C	[6] –	[7] 60	[8] 80	140
Demand	190	140	80	410

Find the optimal solution for this table, starting from the initial allocation shown, and state the minimum cost (unit costs are in £).

Calculate the shadow costs and improvement indices:

	$K_1 = 2$	$K_2 = 4$	$K_3 = 5$
$R_1 = 0$	[2] X	[5] (1)	[1] (-4)
$R_2 = 2$	[4] X	[6] X	[5] (-2)
$R_3 = 3$	[6] (1)	[7] X	[8] X

Improve the solution using the stepping-stones method, using cell $(1,3)$ as the entering cell. There are two exiting cells: $(2,2)$ and $(3,3)$.

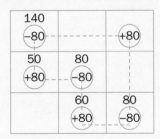

Here is the revised allocation:

From \ To	X	Y	Z
A	[2] 60	[5] –	[1] 80
B	[4] 130	[6] –	[5] –
C	[6] –	[7] 140	[8] –

Try repeating this example using cell $(2,3)$ as the entering cell.

D2

EXAMPLE 2 (CONT.)

This solution is degenerate.

There are four occupied cells whereas $m + n - 1 = 5$

Create an extra occupied cell by inserting a zero in cell $(2, 2)$:

From \ To	X	Y	Z
A	2 60	5 –	1 80
B	4 130	6 0	5 –
C	6 –	7 140	8 –

Recalculate the shadow costs and improvement indices:

	$K_1 = 2$	$K_2 = 4$	$K_3 = 1$
$R_1 = 0$	2 X	5 ①	1 X
$R_2 = 2$	4 X	6 X	5 ②
$R_3 = 3$	6 ①	7 X	8 ④

The solution is optimal.

There are no negative improvement indices.

The minimum cost solution sends 60 units from A to X, 80 units from A to Z, 130 units from B to X and 140 units from C to Y.

Total cost $= 60 \times 2 + 80 \times 1 + 130 \times 4 + 140 \times 7 = £1700$

D2

Exercise 2.5

1

From \ To	X	Y	Z	Supply
A	3	1	4	50
B	2	4	3	40
C	1	5	2	80
Demand	60	30	80	170

a Use the north-west corner method to obtain an initial solution to this transportation problem.

b Explain why this initial solution is degenerate.

c Use the stepping-stones method to obtain an optimal solution and calculate its cost (unit costs are in £).

2

To From	X	Y	Z	Supply
A	9	8	11	150
B	8	6	10	160
C	4	4	7	150
Demand	110	200	150	460

a Use the north-west corner method to find an initial allocation for the transportation problem shown.

b Explain why the solution obtained is degenerate.

c Explain why creating an extra occupied cell at $(2, 1)$ will not enable you to calculate shadow costs.

d Create an extra occupied cell at $(3, 1)$ and hence find an optimal solution to the problem. State the entering and exiting cells used in your method.

e Calculate the cost of your optimal solution.

3

To From	X	Y	Z	Supply
A	7	5	8	80
B	9	6	9	100
C	3	2	4	140
D	5	2	3	70
Demand	50	180	160	390

a Use the north-west corner method to find an initial allocation for the transportation problem shown.

b Use the stepping-stones method to obtain an improved solution.

c Explain why this improved solution is degenerate.

d Calculate improvement indices to show that the solution is optimal.

D2

4

From \ To	X	Y	Z	Supply
A	6	4	7	90
B	8	5	6	95
Demand	70	90	105	

The table shows a situation in which the demand from the three destinations X, Y and Z is greater than the available supply from sources A and B.

a Introduce a dummy source to create a balanced problem.
Use the north-west corner method to find an initial allocation.

b Use the stepping stones method to obtain an improved solution, stating the entering and exiting cells used.

c Explain why this improved solution is degenerate.

d Show that the solution is optimal.

e Explain the implications of the solution for the three destinations X, Y and Z.

5

From \ To	P	Q	R	S	Supply
A	2	5	3	4	80
B	2	4	1	4	50
C	5	6	2	3	100
D	3	7	5	5	30
Demand	60	70	100	30	260

a Show that for the table shown the initial allocation obtained using the north-west corner method is degenerate.

b Obtain an optimal solution to the problem.

c State, with reasons, whether the solution you have obtained is unique.

In a transportation problem you minimise the cost (the objective function) by choosing the values in the cells (the decision variables). This is a linear programming situation, but the particular nature of the problem means you can solve it with the methods described in the previous sections.

It is sometimes useful to express the problem in linear programming terms.

Use x_{ij} to refer to the value allocated to cell (i,j).

It is the same with many other decision maths topics.

E.g. You can treat finding the shortest path through a network as a linear programming problem, but Dijkstra's algorithm allows it to be solved more easily.

EXAMPLE 1

The table shows the demand from two destinations, X and Y, and the supply available from two sources, A and B. Express the problem of finding the least cost transportation allocation as a linear programming formulation.

From \ To	X	Y	Supply
A	3	5	180
B	4	2	120
Demand	130	170	300

The decision variables are x_{11}, x_{12}, x_{21} and x_{22}.

For example, x_{12} is the amount to be supplied by source A to destination Y.

The constraints are provided by the row and column totals. The amount supplied by source A is 180,

so $x_{11} + x_{12} = 180$

Similarly $x_{21} + x_{22} = 120$
$x_{11} + x_{21} = 130$
$x_{12} + x_{22} = 170$

The constraints are equalities because the problem is balanced. An unbalanced problem would require inequalities.

The usual non-negativity constraints apply,

so $x_{11} \geqslant 0, x_{12} \geqslant 0, x_{21} \geqslant 0, x_{22} \geqslant 0$

A balanced problem also means that there are really only three constraints here, as values satisfying the first three equations would automatically satisfy the fourth.

You need to minimise the cost, which is given by

$$C = 3x_{11} + 5x_{12} + 4x_{21} + 2x_{22}$$

The full linear programming formulation is:

Minimise $\qquad C = 3x_{11} + 5x_{12} + 4x_{21} + 2x_{22}$

subject to $\quad x_{11} + x_{12} = 180$
$x_{21} + x_{22} = 120$
$x_{11} + x_{21} = 130$
$x_{12} + x_{22} = 170$
$x_{11} \geqslant 0, x_{12} \geqslant 0, x_{21} \geqslant 0, x_{22} \geqslant 0$

In the examination you will only be required to set up the linear programming formulation. You will **not** be asked to solve the problem in this form.

Exercise 2.6

1 Suppliers A and B carry stocks of cement. On a certain day supplier A has 540 bags in stock, and supplier B has 480 bags. Two building projects, C and D, require 390 and 630 bags respectively.

The transportation costs, in pence per bag, are shown in the table.

From \ To	C	D
A	25	18
B	16	23

Express the problem of finding the minimum cost allocation as a linear programming formulation.

2 The table shows the number of units of a commodity available from three suppliers A, B and C, and the number of units needed by three customers X, Y and Z, together with the transport costs in £ per unit.

From \ To	X	Y	Z	Supply
A	4	2	5	250
B	3	1	4	180
C	6	5	7	140
Demand	200	220	150	570

a Express the problem of finding the optimal allocation as a linear programming formulation.

b Explain why there are only five independent constraints (apart from the non-negativity constraints).

3

From \ To	X	Y	Z	Supply
A	2	3	2	240
B	4	2	3	200
C	1	5	4	260
Demand	180	210	220	

Express the problem of finding the optimal allocation for the transportation tableau shown as a linear programming formulation.

D2

1 a Use the north-west corner method to find an initial allocation for this transportation tableau.
Calculate the total cost of this allocation.

b Calculate the improvement indices for your allocation, and explain how they show that the allocation is not optimal.

c Use the stepping-stones method to find an improved allocation, and show that it is optimal. Calculate the cost of this optimal solution.

From \ To	X	Y	Z	Supply
A	3	5	4	160
B	4	3	2	250
C	5	6	5	200
Demand	190	190	230	610

2 A car hire company keeps its cars at three garages, A, B and C, from where they are supplied to three hire outlets, X, Y and Z. The table shows the numbers of cars available and needed on a certain day, together with the cost, in £ per car, of getting the cars from the garages to the outlets.

From \ To	X	Y	Z	Cars available
A	£15	£20	£24	12
B	£12	£16	£10	9
C	£28	£25	£25	8
Cars needed	8	6	10	

a Explain why it is necessary to add an extra column to the table in order to find an allocation.

b Use the north-west corner rule to find an initial allocation.

c Calculate shadow costs and improvement indices.
Hence find an improved allocation.
State the entering and exiting cells you have used.

d Show that the new allocation is optimal and find its cost.

e Explain the meaning of the value remaining in the extra column.

3 A theatre club wishes to book tickets for a play which is being performed on Thursday, Friday and Saturday. They need 40 standard tickets and 25 senior citizen tickets. Unfortunately, there are only 30 seats available for the Thursday performance, 28 for the Friday and 22 for the Saturday. The table shows the prices of tickets for the three performances.

	Standard	Senior Citizen
Thursday	£10	£6
Friday	£15	£12
Saturday	£18	£12

a Draw up a transportation tableau suitable for solving this problem (notice that 65 seats are required and 80 seats are available).

b Find an initial allocation using the north-west corner method.

c Show that the allocation you have obtained is not optimal.

d Use the stepping-stones method to find an improved allocation, and show that it is optimal.

4 a Use the north-west corner method to find an initial allocation for the transportation problem shown in the table.

b Explain why the solution you have found is degenerate.

c By creating an extra occupied cell, calculate improvement indices and show that the solution is not yet optimal.

d Use the stepping-stones method to find the optimal solution, stating the entering and exiting cells used.

From \ To	X	Y	Z	Supply
A	5	3	2	45
B	7	6	3	42
C	9	6	2	18
Demand	60	27	18	105

5 Express the transportation problem shown in question **1** of this exercise as a linear programming formulation.

D2

Summary

Refer to

○ You can find an initial allocation for a transportation problem
using the north-west corner method with a transportation tableau.   2.1

○ Cell (i, j) is in the ith row and the jth column, and has unit cost C_{ij}.
You test and improve the solution as follows:
   ○ Calculate shadow costs R_i for rows and K_j for columns so that
   for each occupied cell (i, j) $R_i + K_j = C_{ij}$
   ○ Calculate improvement indices I_{ij} for each unoccupied cell (i, j),
   where $I_{ij} = C_{ij} - R_i - K_j$
   ○ An allocation is optimal if and only if $I_{ij} \geqslant 0$ for all unoccupied cells.   2.2

○ You improve a non-optimal solution by allocating units to a cell
with a negative improvement index. This requires a sequence of
adjustments around a loop, and is called the stepping-stones method.   2.3

○ The problem is balanced if total supply equals total demand.
If the problem is unbalanced you add a dummy row or column,
with zero unit costs, to balance the problem.   2.4

○ If a table with m rows and n columns has fewer than $(m + n - 1)$
occupied cells the solution is degenerate. In such cases you create
extra occupied cells by inserting zeros.   2.5

○ You can express a transportation problem as a linear programming formulation.   2.6

Links

Large-scale problems of the type described
in this chapter are solved by oil companies
in the different stages of the distribution
of their products.

Crude oil is transported from oil fields to
refineries, then petroleum is transported
from the refineries to depots. Finally, the
petroleum is dispatched to the retailers.

The distances of the journeys involved
and the quantities of the substances being
transported make it important that the
various stages of transportation are as
efficient as possible.

D2

3

Allocation (assignment) problems

This chapter will enable you to
- create an opportunity cost matrix for an assignment problem
- test an opportunity cost matrix for an optimal assignment
- revise an opportunity cost matrix
- express an assignment problem as a linear programming formulation.

Introduction

In matching problems, members of one set (e.g. workers)
are paired with members of another set (e.g. tasks).

In this chapter you will deal with situations in
which all pairings are possible but each has an
associated cost. The aim is to decide on the pairing
which has the least total cost.

Suppose workers *A*, *B* and *C* are to be assigned to tasks *X*, *Y* and *Z*. They provide estimated times (in hours) for completing the tasks as shown in this table:

	X	Y	Z
A	7	10	6
B	9	9	7
C	12	8	10

You could treat this as a transportation problem in which each worker supplies 1 unit and each task demands 1 unit, but this is a very inefficient approach.

This is a cost matrix. You need to assign one worker to each task so that the total number of hours involved is a minimum.

Use the Hungarian algorithm as follows:

Step 1 Construct an opportunity cost matrix to show the relative costs of assignments.
Step 2 Test if an optimal assignment can be made. If so, make the assignment and stop.
Step 3 Revise the matrix and return to Step 2.

To construct an opportunity cost matrix use the following fact:

The optimal assignment is not changed if all of the costs in a given row (or column) are changed by the same amount.

For example, if worker *A* increased her estimated times to 10, 13 and 9 hours, you would assign workers to tasks in the same order, although the total time would increase by 3 hours.

Following this rule, subtract the smallest number in each row from all the numbers in that row (this is called row reduction). The table becomes

	X	Y	Z
A	1	4	0
B	2	2	0
C	4	0	2

The zero(s) generated in each row by row reduction correspond to the best task(s) for that worker.

Subtract the smallest number in each column from all the numbers in that column (this is called column reduction).

The result is an opportunity cost matrix:

	X	Y	Z
A	0	4	0
B	1	2	0
C	3	0	2

The zero(s) generated in each column by column reduction correspond to the best worker(s) for that task.

If possible assign workers to tasks using only cells containing zeros. In this case, task X must be done by worker A and task Y by worker C. This leaves worker B to do task Z.

Look at the original table to find the total cost:

Worker A – task X 7 hours
Worker B – task Z 7 hours
Worker C – task Y 8 hours
           Total cost   22 hours

> You will be expected to reduce rows first, then columns. Reducing columns first would produce a different opportunity cost matrix, but you would arrive at the same assignment.

> You will meet the general method of testing whether an assignment is possible in Section 3.2.

Exercise 3.1

1 For each of the following cost matrices perform a row and a column reduction to produce an opportunity cost matrix. If possible, make an assignment and find its total cost.

a

	X	Y	Z
A	8	2	7
B	6	4	10
C	8	2	1

b

	P	Q	R	S
A	14	17	8	13
B	20	27	23	21
C	15	30	18	14
D	19	31	24	20

c

	X	Y	Z
A	19	9	13
B	26	14	17
C	26	15	11

d

	P	Q	R	S
A	12	8	13	11
B	11	9	7	4
C	7	5	9	11
D	6	9	13	13

2 For the given cost matrix, generate opportunity cost matrices by

   a   reducing rows then columns

   b   reducing columns then rows.

   c   Show that an assignment can be made in case **a** but that no assignment is possible in case **b** without revising the matrix further.

	P	Q	R	S
A	11	15	15	12
B	13	16	10	16
C	12	11	16	7
D	14	19	13	13

D2

Here are two opportunity cost matrices, each with seven zeros:

	P	Q	R	S
A	0	0	2	6
B	5	1	0	2
C	4	0	6	0
D	1	4	0	0

An assignment can be made as shown.

	P	Q	R	S
A	0	2	0	3
B	1	3	0	5
C	2	0	6	0
D	0	4	0	1

No assignment can be made.

	P	Q	R	S
A	0	0	2	6
B	5	1	0	2
C	4	0	6	0
D	1	4	0	0

You need at least four horizontal and/or vertical lines to cover all of the zeros.

	P	Q	R	S
A	0	2	0	3
B	1	3	0	5
C	2	0	6	0
D	0	4	0	1

You can cover all of the zeros with only three horizontal and/or vertical lines.

If the minimum number of lines needed to cover all the zeros in an $n \times n$ opportunity cost matrix is m, then

an assignment can be made if $m = n$

For a very large matrix you would need an algorithm to find this minimum configuration of lines, but for small matrices you can find it by inspection.

If an assignment cannot be made, revise the matrix to create zeros in different places, as follows:

- Draw the minimum set of lines needed to cover all the zeros.
- Identify the smallest uncovered number, s.
- Subtract s from all the uncovered numbers.
- Add s to any number at the intersection of two lines.

This is equivalent to subtracting s throughout each uncovered row and adding s throughout each covered column, so the optimal assignment is unchanged.

For the second case shown, the smallest uncovered number is 1. Subtract this from all the uncovered numbers, and add it to cell(3, 1) and cell(3, 3).

The zeros in this revised matrix now require four lines to cover them, as shown. An optimal assignment is now possible, and is shown highlighted.

	P	Q	R	S
A	0	1	0	2
B	1	2	0	4
C	3	0	7	0
D	0	3	0	0

D2

EXAMPLE 1

This table shows the costs, in £thousand, quoted by five building companies A, B, C, D and E, to complete five projects P, Q, R, S and T. Each company is to be assigned one project. Find the optimal assignment and calculate its cost.

	P	Q	R	S	T
A	42	37	48	42	40
B	56	40	39	52	45
C	45	60	36	59	48
D	51	48	46	52	54
E	37	41	49	49	44

First carry out a row reduction:

	P	Q	R	S	T
A	5	0	11	5	3
B	17	1	0	13	6
C	9	24	0	23	12
D	5	2	0	6	8
E	0	4	12	12	7

A 5 × 5 table is the largest you will meet in the examination.

Subtract 37 from the first row, 39 from the second row and so on.

Carry out a column reduction, giving an opportunity cost matrix:

The zeros in this matrix can be covered using three lines, as shown, so you need to revise the matrix.

	P	Q	R	S	T
A	5	0	11	0	0
B	17	1	0	8	3
C	9	24	0	18	9
D	5	2	0	1	5
E	0	4	12	7	4

Subtract 5 from the fourth column and 3 from the fifth column. Columns containing zeros will not change.

The smallest uncovered number is 1. Subtract this from all the uncovered numbers, and add it to cell(1,3) and cell(5,3).

The revised matrix now requires five lines to cover the zeros, as shown, so an optimal assignment is possible.

	P	Q	R	S	T
A	5	0	12	0	0
B	16	0	0	7	2
C	8	23	0	17	8
D	5	1	0	0	4
E	0	4	13	7	4

You may need to revise the matrix more than once before an assignment can be made.

First look for forced choices:

Task P can only be done by E.
Task T can only be done by A.
Company C can only do task R.
This leaves B to do task Q and D to do task S, as shown.

	P	Q	R	S	T
A	5	0	12	0	0
B	16	0	0	7	2
C	8	23	0	17	8
D	5	1	0	0	4
E	0	4	13	7	4

More than one pattern of assignments may be possible (though not in this case), but they will all give the same total cost.

Example 1 is continued on the next page.

D2

The original table gives
a total cost of 37 + 40 + 36 + 52 + 40 = 205
so the five projects will
cost a total of £205 000.

	P	Q	R	S	T
A	42	37	48	42	40
B	56	40	39	52	45
C	45	60	36	59	48
D	51	48	46	52	54
E	37	41	49	49	44

Exercise 3.2

1 For each of these opportunity cost matrices, decide if optimal
assignment(s) can be made. Make the assignments if possible.

a

	P	Q	R	S
A	4	0	3	1
B	2	0	5	0
C	1	3	2	0
D	6	0	0	3

b

	P	Q	R	S
A	8	5	0	0
B	0	3	4	7
C	3	0	0	2
D	9	5	0	7

c

	P	Q	R	S	T
A	0	7	0	2	0
B	4	0	8	8	2
C	0	3	3	0	4
D	5	0	2	0	9
E	1	0	3	5	2

d

	P	Q	R	S	T
A	0	2	5	4	7
B	1	1	0	3	4
C	1	0	2	0	8
D	3	0	3	2	0
E	9	6	6	0	0

2 The table shows the personal best length
times, in seconds, of each of four swimmers
in a medley relay team.

a Construct an opportunity cost matrix.

b Revise the matrix as necessary until an
assignment can be made.

c List the two possible optimal assignments.

d If every swimmer equals his personal best,
what will be the team's time for the race?

	Front crawl	Breast stroke	Back crawl	Butterfly
George	43	56	48	52
Harry	49	53	50	55
Ian	41	55	47	50
John	44	59	47	53

3 The bipartite graph shows the price, in hundreds of pounds, quoted by three building firms *A*, *B* and *C* for each of three projects. Each firm is to be assigned one project.

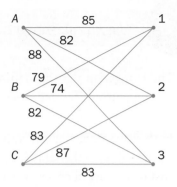

a Show these data as a cost matrix.

b Construct an opportunity cost matrix.

c By revising your matrix if necessary, assign the projects and find the total cost incurred.

4 A company opens four branches at *P*, *Q*, *R* and *S* on this network. The four managers chosen to run these branches live at *A*, *B*, *C* and *D*. The weights correspond to distances, in kilometres.

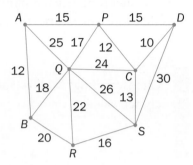

a Draw up a table showing the minimum travelling distances of the managers to the branches (find these by inspection).

b Construct an opportunity cost matrix. Show that an optimal assignment cannot yet be made.

c Revise the matrix and hence find the assignment of managers to branches which minimises the total travelling distance.

5 Five teams of workers, *A*, *B*, *C*, *D* and *E*, are to be assigned to five projects. The estimated completion times, in days, are shown in the table.

Find the optimal assignment of teams to projects and state the expected total number of days involved.

	P	Q	R	S	T
A	9	15	6	20	10
B	7	15	11	15	12
C	12	14	11	22	12
D	8	17	9	18	14
E	10	13	8	19	14

6 Four developers, *A*, *B*, *C* and *D*, are invited to tender for each of four development contracts. They are each only equipped to work on one development. The table shows the bids in £thousands. The * indicates that a bid was not made.

a Treating the * as an infinitely high cost, set up an opportunity cost matrix.

b Show that there is just one way in which the contracts could be allocated to minimise the development cost.

c Calculate the total cost.

	1	2	3	4
A	125	220	160	85
B	140	180	160	105
C	130	200	*	95
D	135	210	170	100

D2

The Hungarian algorithm minimises the total value of the chosen pairings. Sometimes you want to maximise the total – for example, the table might show profits arising from the pairings.

A table of potential gains is sometimes called a **payoff matrix**.

In this situation, you can construct a cost matrix by subtracting all the values in the table from the largest value. This shows the cost or penalty of choosing each cell.

The assignment which minimises this cost will be the same as that which maximises the payoff in the original table.

For example, for a table with a maximum value of 50, choosing a cell with a value of 40 would 'cost' you 10. You could in fact subtract them all from any chosen value greater than or equal to the largest value in the table.

EXAMPLE 1

Choose pairings for this table to maximise the total of the chosen values.

	1	2	3	4
A	23	34	27	25
B	33	29	16	20
C	47	31	36	34
D	28	40	30	35

Construct a cost matrix.
The largest value in the table is 47.

Subtract each value from 47:

	1	2	3	4
A	24	13	20	22
B	14	18	31	27
C	0	16	11	13
D	19	7	17	12

Carry out a row reduction:

	1	2	3	4
A	11	0	7	9
B	0	4	17	13
C	0	16	11	13
D	12	0	10	5

Carry out a column reduction:

You can cover the zeros with three lines, so an optimal assignment cannot be made.

	1	2	3	4
A	11	0	0	4
B	0	4	10	8
C	0	16	4	8
D	12	0	3	0

The smallest uncovered number is 4.

D2

60

EXAMPLE 1 (CONT.)

Revise the matrix by subtracting 4 from all uncovered cells and adding 4 to cell $(1, 1)$ and cell $(4, 1)$:

Four lines are now needed, so an assignment can be made.

	1	2	3	4
A	15	0	0	4
B	0	0	6	4
C	0	12	0	4
D	16	0	3	0

	1	2	3	4
A	15	0	0	4
B	0	0	6	4
C	0	12	0	4
D	16	0	3	0

	1	2	3	4
A	15	0	0	4
B	0	0	6	4
C	0	12	0	4
D	16	0	3	0

There are two possible assignments, as shown. Relating these to the original table, you have

$$\{A3, B2, C1, D4\} = 27 + 29 + 47 + 35 = 138$$
or $$\{A2, B1, C3, D4\} = 34 + 33 + 36 + 35 = 138$$

Exercise 3.3

1 Five employees are to be placed in five roles, A–E, within an organisation. Before the appointments are made, the employees take aptitude tests for the various roles. The scores they obtain are shown in the table. The appointments are to be made so as to maximise the total test score.

	A	B	C	D	E
Ms Pugh	35	49	38	47	44
Mr Quentin	31	45	43	45	29
Dr Rani	47	49	46	55	38
Mrs Stuart	40	48	55	59	43
Miss Till	51	55	56	67	45

a Construct a corresponding cost matrix for this maximisation problem.

b Construct an opportunity cost matrix.

c By revising your matrix, assign the employees in order to maximise the total test score.

2 Lisa, Marilyn, Nicola and Olive are each to be entered in one of four throwing events – shot, javelin, discus and hammer. The team score will be the total of the distances (in metres) they achieve. The table shows their best throws in a recent training session.

	Shot	Javelin	Discus	Hammer
Lisa	12	49	40	37
Marilyn	10	50	41	34
Nicola	14	46	38	36
Olive	13	54	45	30

a Set up an opportunity cost matrix.

b Revise your matrix to show that there are two ways in which the entries could be made to maximise the expected total distance.

c Calculate the expected team score.

D2

A problem is **unbalanced** if the numbers of items in the two sets to be matched are different. You can overcome the difficulty by adding dummy row(s) or column(s).

This is related to the unbalanced transportation problems you met in Section 2.4.

The values placed in a dummy row or column must all be equal. It is usual to make them all zero.

EXAMPLE 1

Five firms, A, B, C, D and E, tender for four jobs, 1–4. The bids offered, in £thousand, are shown in the table. Find the least cost allocation of the jobs.

	1	2	3	4
A	7	9	10	7
B	12	15	11	10
C	12	10	13	14
D	8	5	9	8
E	10	13	9	12

An extra column is needed to balance the problem.

Add a dummy column of zeros:

	1	2	3	4	Dummy
A	7	9	10	7	0
B	12	15	11	10	0
C	12	10	13	14	0
D	8	5	9	8	0
E	10	13	9	12	0

A row reduction will not alter the matrix.

Carry out a column reduction: The zeros can be covered by four lines, so revise the matrix.

	1	2	3	4	Dummy
A	0	4	1	0	0
B	5	10	2	3	0
C	5	5	4	7	0
D	1	0	0	1	0
E	3	8	0	5	0

Subtract 3 from all uncovered cells and add 3 to cells $(1,3)$, $(1,5)$, $(4,3)$ and $(4,5)$:

Five lines are now needed, so you can make an assignment.

	1	2	3	4	Dummy
A	0	4	4	0	3
B	2	7	2	0	0
C	2	2	4	4	0
D	1	0	3	1	3
E	0	5	0	2	0

Allocate job 1 to firm A, job 4 to B, job 2 to D and job 3 to E. Firm C will not be offered a job.

	1	2	3	4	Dummy
A	0	4	4	0	3
B	2	7	2	0	0
C	2	2	4	4	0
D	1	0	3	1	3
E	0	5	0	2	0

Exercise 3.4

1   Four language teachers have each taught A level French, German and Spanish. The percentage of their students failing on the last occasion they taught each language is shown in the table.

Based on this evidence, which language should be taught by which teacher in the coming year?

Language Teacher	French	German	Spanish
Mrs Atkins	7.2	6.4	10.4
Mr Beaumont	8.5	9.6	9.2
Mr Caldwell	12.2	7.5	7.8
Mr Delgado	8.8	9.3	7

2   An employment bureau has four workers available and has been asked to supply one worker at each of five locations. The table shows the distance in miles that each worker would have to travel to each location.

Location Worker	1	2	3	4	5
A	8	12	10	7	11
B	4	7	11	12	6
C	5	6	8	10	8
D	9	11	6	8	14

a   Construct an opportunity cost matrix. Show that an optimal allocation cannot yet be made.

b   Revise the matrix and hence allocate the workers to the locations.

c   The distances shown are one-way and the bureau pays £0.40 per mile in travel expenses. Calculate the total travel bill for a five-day week.

3   A local newspaper runs a competition, the prizes being a pair of tickets to each of five theatrical productions.

One winner opts to take the money instead. The remaining four are asked to rate the productions on a scale of 1 (not keen) to 10 (must see), and the prizes are allocated so as to maximise the total approval rating.

Production Winner	1	2	3	4	5
A	3	8	5	6	5
B	8	4	5	7	4
C	9	7	7	10	6
D	6	8	4	8	6

a   Construct a cost matrix for this maximisation problem.

b   Find an opportunity cost matrix and revise this as necessary to enable an optimal assignment to be made.

c   Which theatre production will not be attended by a prize winner?

D2

You can express an assignment problem in linear programming terms.

You can think of the problem as attaching a 1 (assigned) or a 0 (not assigned) to each cell of the table.

> The values attached to the cells are the decision variables.

You need exactly one 1 in each row and column.
The row and column totals are therefore all equal to 1.

> The row and column totals provide the constraints.

If you multiply the cost of each cell by the 1 or 0 attached to it and add, you get the total cost of the whole allocation.
You need to minimise this cost.

EXAMPLE 1

The table shows the times, in hours, that three workers, A, B and C, estimate they will take to do tasks P, Q and R. The tasks are to be assigned so as to minimise the total time. Express this as a linear programming problem.

	P	Q	R
A	15	12	8
B	10	13	11
C	20	16	18

Each cell (i, j) is given an associated value $x_{ij} = 1$ or 0.
Each worker does only one task, so

$$x_{11} + x_{12} + x_{13} = 1 \qquad [1]$$
$$x_{21} + x_{22} + x_{23} = 1 \qquad [2]$$
$$x_{31} + x_{32} + x_{33} = 1 \qquad [3]$$

Each task is done by only one worker, so

$$x_{11} + x_{21} + x_{31} = 1 \qquad [4]$$
$$x_{12} + x_{22} + x_{32} = 1 \qquad [5]$$
$$x_{13} + x_{23} + x_{33} = 1 \qquad [6]$$

> Not all of these constraints are really needed, because if the first five are satisfied the sixth will automatically be true. (In fact, you can get equation [6] by adding [1], [2] and [3] and subtracting [4] and [5].) There are five **independent** constraints.

The total time taken is:
$$T = 15x_{11} + 12x_{12} + 8x_{13} + 10x_{21} + 13x_{22} + 11x_{23} + 20x_{31} + 16x_{32} + 18x_{33}$$

The problem is therefore to minimise T subject to the constraints [1]–[6]. There is no need for non-negativity constraints because x_{ij} is defined as being either 1 or 0.

> You will **not** be expected to solve assignment problems by linear programming methods for this examination, but you may be asked to formulate the problem in those terms.

Exercise 3.5

1 The table shows the price quoted (in £hundreds) by
builders *A*, *B* and *C* for three contracts. Each builder is
to be awarded one contract, which will be allocated so as
to minimise the total cost of the three jobs. The problem
is to be solved by linear programming methods.

	1	2	3
A	5	12	7
B	6	11	6
C	10	14	9

    a State the decision variables and explain the meaning
of the values they can take.

    b List the constraints.

    c State the objective function.

2 Workers *A*, *B* and *C* are to be allocated one of three
jobs *P*, *Q* and *R* within an organisation. They undergo
psychometric testing which scores their suitability for
the jobs, as shown in the table. The jobs are then
allocated so as to maximise the total test score.

	P	*Q*	*R*
A	86	72	67
B	58	66	60
C	74	76	80

Express this problem as a linear programming formulation.

3 An assignment problem involves matching five workers to five jobs.
It is to be expressed as a linear programming formulation.

    a How many decision variables are there?

    b How many constraints are there? How many of these
are independent?

1 Bob, Carol and Deepak need to take taxis to their homes. There are three taxis waiting at the rank. The fares they quote for the three journeys are shown in the table.

	Bob	Carol	Deepak
Luxicabs	£7.50	£5.00	£10.00
Maxitaxi	£8.50	£4.50	£10.50
Nobby's	£10.00	£4.50	£10.00

    a Construct an opportunity cost matrix.

    b Explain how you know that an optimal assignment cannot yet be made.

    c Revise the matrix until an assignment is possible.

    d State which person should travel in which taxi, and calculate the total cost of the trips.

2 Five people, *A*, *B*, *C*, *D* and *E*, have made sealed bids to buy one of four houses, 1–4. The table shows the value, in tens of thousands of pounds, of the bids. The houses will be allocated to maximise the total money paid.

	1	2	3	4
A	23	16	22	18
B	18	12	25	15
C	20	16	25	17
D	16	17	24	18
E	21	18	20	20

    a Explain why it is necessary to add an extra column to the table.

    b Construct a cost matrix and from it obtain an opportunity cost matrix.

    c Revise the matrix until an optimal assignment can be made.

    d Show that either *B* or *C* will not get a house.

    e Calculate the total money paid for the four houses.

3 Express the problem described in question **1** as a linear programming formulation.

4 As part of a regeneration project the government asks four local councils to devise schemes under four headings. They must indicate how much money each scheme will need and how many jobs they hope to create. Each council will be awarded one of the schemes.

Table 1 shows the projected cost, in £millions, of each scheme. Table 2 shows the estimated number of jobs created.

Scheme \ Council	A	B	C	D
1	8.5	6.6	7.1	10.4
2	8.4	6.9	7.8	9.6
3	7.9	6.2	8.6	11.5
4	8.9	6.6	7.5	10.8

Table 1

a Suppose the schemes are awarded so that the total cost is minimised.

  i Find the allocation of schemes to councils which would achieve this.
  ii Find the cost per job created under this plan.

b It is decided instead that the schemes will be awarded to maximise the number of jobs created.

Scheme \ Council	A	B	C	D
1	75	63	70	80
2	72	73	83	78
3	76	62	71	81
4	70	60	71	77

Table 2

  i Show that there are two ways in which this allocation can be made.
  ii Find the cost per job created under the cheaper of these two allocations.

5 Hortense has a vegetable garden divided into five plots. She is going to plant a different vegetable in each of four plots, and erect a shed on the fifth.

She tests the soil in each plot and calculates a score for its suitability for each of the four vegetables. Her results are shown in the table. She wishes to maximise the scores for the four plots planted.

Vegetable \ Plot	1	2	3	4	5
A	14	10	6	16	12
B	18	13	6	14	10
C	7	8	10	15	10
D	16	9	15	13	17

a Find a suitable allocation of vegetables to plots.

b Where will she put her shed?

D2

Summary

Refer to

- The Hungarian algorithm minimises total costs.
- To obtain an opportunity cost matrix you carry out
  - a row reduction
    - subtract the minimum value in each row from every cell in the row
  - a column reduction
    - subtract the minimum value in each column from every cell in that column.

    3.1
- To test and revise an $n \times n$ opportunity cost matrix
  - Draw the minimum number (m) of lines need to cover all the zeros
  - If $m = n$, an optimal assignment can be made.
    Make the assignment and stop
  - Identify the smallest uncovered number, s
  - Subtract s from all the uncovered numbers
  - Add s to any number at the intersection of two lines.

    3.2
- To maximise the total gain, convert the payoff matrix to a
  cost matrix by subtracting each value from the largest value
  in the table. Minimising the resulting costs is the same as
  maximising the gain.

  3.3
- You can convert an unbalanced problem to a balanced problem
  by adding dummy row(s) or column(s) containing equal values,
  usually zeros.

  3.4
- To express the problem in linear programming form
  - The decision variable for cell (i, j) is $x_{ij} = 1$ (assigned) or 0 (not assigned)
  - The constraints are that the sum of $x_{ij} = 1$ for each row and column
  - The objective function is $C = $ sum of $(x_{ij} \times$ cost of cell $(i, j))$

    3.5

Links

The Hungarian algorithm can be used to achieve
efficient transport planning in a city with multiple
vehicles available.

In a city with a public transport system consisting
of buses, trains and an underground train system,
the Hungarian algorithm could be used to ensure
that all necessary journeys were covered by the
combined timetables in the most efficient way.

D2

4

Game theory

This chapter will enable you to
- analyse a two-person zero-sum game
- determine if a game has a stable solution
- use dominance to simplify a payoff matrix
- find a mixed strategy for games with no stable solution
- express a game as a linear programming formulation.

Introduction

Game theory is a recent development in mathematics, having only been studied extensively since the 1940s. It deals with the analysis of competitive situations and has been applied in a wide variety of commercial and military fields.

This chapter looks only at two-person games where one side's gain exactly equals the other side's loss (zero-sum games).

Problems with more than two competitors or where both/all sides may gain or lose (non zero-sum games) are still poorly understood and there is much ongoing research.

Ann and Bob each play a card – king (*K*),
queen (*Q*) or jack (*J*).
If both play *K*, Bob pays Ann 5p.
If Ann plays *K* and Bob plays *Q*, Ann pays Bob 4p.
You could record these as (5,–5) and (–4,4).

(5,–5) means Ann gains
5 and Bob loses 5.

This table shows the complete
set of payments.

This is a two-person game
– there are just two competitors.

		Bob		
		K	**Q**	**J**
Ann	**K**	(5,–5)	(–4,4)	(2,–2)
	Q	(3,–3)	(1,–1)	(4,–4)
	J	(2,–2)	(3,–3)	(–1,1)

The game could involve
two teams rather than
two people.

On each play, one competitor's gain equals the other's loss.
It is a zero-sum game.

> In a zero-sum game, the sum of the competitors' gains and
> losses on each play is zero.

D2

For a zero-sum game you only need to record one number in each
cell, as the other is just its negative. The table becomes:

		Bob		
		K	**Q**	**J**
Ann	**K**	5	–4	2
	Q	3	1	4
	J	2	3	–1

This is Ann's payoff matrix.

Analyse the game from Ann's point of view:

		Bob			Row minimum
		K	**Q**	**J**	
Ann	**K**	5	–4	2	–4
	Q	3	1	4	1	← max = 1
	J	2	3	–1	–1

If she plays *K*, the
worst outcome is
that she loses 4p.

If she plays *Q*,
the worst is that
she wins 1p.

If she plays *J*, the
worst is that she
loses 1p.

It is the convention to record the
gains of the player on the left of
the table – the **row player**.

Bob's payoff matrix would look
like this:

		Ann		
		K	**Q**	**J**
Bob	**K**	–5	–3	–2
	Q	4	–1	–3
	J	–2	–4	1

The worst outcomes for Ann are
the row minimums.

Her **play safe** strategy is to play Q, which gives the best guaranteed outcome (win 1p). This is a **maximin** strategy, because it maximises her minimum gain.

> A play safe strategy gives the best guaranteed outcome regardless of what the other player does.

On the same table analyse the game from Bob's point of view.

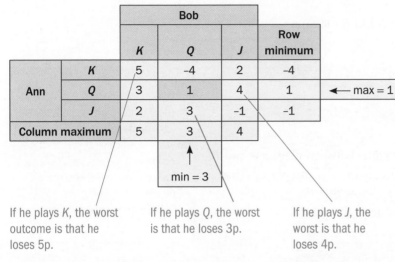

		Bob			Row minimum
		K	Q	J	
Ann	K	5	–4	2	–4
	Q	3	1	4	1 ← max = 1
	J	2	3	–1	–1
Column maximum		5	3	4	

min = 3

The worst outcomes for Bob are the column maximums.

If he plays K, the worst outcome is that he loses 5p.

If he plays Q, the worst is that he loses 3p.

If he plays J, the worst is that he loses 4p.

His play safe strategy is to play Q, which gives the best guaranteed outcome (lose 3p). This is a **minimax** strategy, because it minimises his maximum loss.

If both players play safe every time, it is a **pure strategy** game. Ann wins 1p each time. This is the **value** of the game to Ann.

The value of the game to Bob is –1p.

If Ann knows Bob will play Q, she should play J instead of Q, winning 3p instead of 1p. This means the solution is **unstable**.

However, if Bob knows Ann will play Q, he should still play Q, because the other options are worse.

> A game has a **stable solution** if neither player can gain by changing from their play safe strategy.

If a player changes strategy on different plays it is a **mixed strategy** game. You will study these in Section 4.3.

Suppose the payoff matrix is changed, as follows:

		Bob			Row minimum
		K	Q	J	
Ann	K	4	2	–4	–4
	Q	3	4	2	2 ← max = 2
	J	2	–2	1	–2
Column maximum		4	4	2	

min = 2

Ann's play safe strategy is still to play Q. Bob's is now to play J.

Neither player can gain by changing strategy – if Ann plays Q Bob's best option is to play J, and if Bob plays J Ann's best option is to play Q. The game has a stable solution. The value of the game (to Ann) is 2p.

> The stated value of a game is always the value to the row player.

> The value of a game is the payoff to the row player if both players use their best strategy.

In this stable solution:

Max of row minima = min of column maxima = 2

> A game has a stable solution if
> max of row minima = min of column maxima

A stable solution is also called a saddle point.

The 2 in the table is the lowest value in its row and the highest value in its column, just as the centre of a horse's saddle is its lowest point in the nose-to-tail direction and its highest point in the side-to-side direction.

EXAMPLE 1

This payoff matrix describes a game between players A and B. Player A has three strategies, A_1, A_2 and A_3, and player B has four strategies, B_1, B_2, B_3 and B_4. Show that the game has a stable solution, and find the value of the game.

		B			
		B_1	B_2	B_3	B_4
A	A_1	8	6	9	6
	A_2	4	3	-2	2
	A_3	10	-1	1	5

Find the row minima and choose the maximum of these.
Find the column maxima and choose the minimum of these.

		B					
		B_1	B_2	B_3	B_4	Row min	
A	A_1	8	6	9	6	6	← max = 6
	A_2	4	3	-2	2	-2	
	A_3	10	-1	1	5	-1	
Column max		10	6	9	6		
			↑ min = 6		↑ min = 6		

A's play safe strategy is A_1.
B's play safe strategy is either B_2 or B_4.

Max of row minima = min of column maxima, so the game has a stable solution. The value of the game is 6.

> A game can have more than one saddle point – in this case (A_1, B_2) and (A_1, B_4).

D2

Exercise 4.1

1 For each of the following payoff matrices, find the play safe strategy for each player. Determine whether the game has a stable solution and, if so, state the value of the game.

a

		B		
		B_1	B_2	B_3
	A_1	5	3	9
A	A_2	4	7	6
	A_3	2	4	5

b

		B		
		B_1	B_2	B_3
	A_1	9	-2	3
A	A_2	-1	-4	2
	A_3	4	6	3

c

		B		
		B_1	B_2	B_3
	A_1	6	-1	4
A	A_2	-1	-3	5

d

		B	
		B_1	B_2
	A_1	-1	3
A	A_2	2	7
	A_3	6	-2

e

		B			
		B_1	B_2	B_3	B_4
	A_1	4	-2	5	9
A	A_2	2	1	3	5
	A_3	3	-1	-8	4

f

		B			
		B_1	B_2	B_3	B_4
	A_1	3	2	7	10
A	A_2	5	5	7	8
	A_3	-3	2	-5	6

2 For the game shown in question **1d**, write down the payoff matrix for player *B*.

3 A wider definition of a zero-sum game is that there is a fixed number of points awarded on each play.
For example, if there are 10 points available in a game between *A* and *B*, possible scores might be $(10, 0)$ or $(4, 6)$. These have the same effect as $(5, -5)$ and $(-1, 1)$ respectively, as in the first case *A* moves 10 points ahead of *B* and in the second case *B* moves 2 points ahead of *A*.

The table shows a game in which a total of 6 points is awarded for each play.

a Construct the conventional payoff matrix equivalent to this game.

b Find the play safe strategy for each player and show that the game has a saddle point.

c If the game is played ten times and each player follows a pure strategy, what is the final score?

		B		
		B_1	B_2	B_3
	A_1	(4,2)	(3,3)	(2,4)
A	A_2	(2,4)	(0,6)	(3,3)
	A_3	(6,0)	(4,2)	(5,1)

D2

In some games one or more strategies should never be used. Those rows or columns can be removed from the payoff matrix.

In this game A would never use strategy A_2 because A_1 has a better payoff in all three columns. Row 1 **dominates** row 2. You can eliminate row 2.

		B		
		B_1	B_2	B_3
A	A_1	7	3	9
	A_2	4	1	6
	A_3	4	4	-1

In the revised matrix B would never use strategy B_1 because its outcomes are the same or worse for B than those of B_2. Column 2 dominates column 1. You can eliminate column 1.

		B		
		B_1	B_2	B_3
A	A_1	7	3	9
	A_3	4	4	-1

There is now no dominance, so the matrix cannot be simplified further.

		B	
		B_2	B_3
A	A_1	3	9
	A_3	4	-1

In a payoff matrix
- row i dominates row j if, for every column,
  value in row $i \geqslant$ value in row j
- column i dominates column j if, for every row,
  value in column $i \leqslant$ value in column j

Remember that small values are better for player B.

EXAMPLE 1

Use dominance to simplify this payoff matrix as far as possible. What can be deduced from the result?

		B		
		B_1	B_2	B_3
A	A_1	5	3	3
	A_2	-2	7	-1

Column 3 dominates column 2, so eliminate column 2.

		B	
		B_1	B_3
A	A_1	5	3
	A_2	-2	-1

Column 3 values $\leqslant$ column 2 values.

Now row 1 dominates row 2, so eliminate row 2.

		B	
		B_1	B_3
A	A_1	5	3

Row 1 values $\geqslant$ row 2 values.

Strategy B_3 dominates B_1. Eliminating this reduces the matrix to a single cell.

		B
		B_3
A	A_1	3

This shows that there is a stable solution. Player A always plays strategy A_1, player B always plays strategy B_3 and the value of the game is 3.

You can always reduce a game with saddle points to a matrix containing just those cells by using dominance.

Exercise 4.2

1 For each of the following payoff matrices, use dominance to simplify the problem as far as possible. If there is a stable solution, state the strategies the players should adopt and the value of the game.

a

A		B_1	B_2
	A_1	-1	2
	A_2	2	3
	A_3	4	-2

(with *B* heading spanning B_1, B_2)

b

A		B_1	B_2	B_3
	A_1	4	-1	4
	A_2	-1	-3	-4

(with *B* heading spanning B_1, B_2, B_3)

c

A		B_1	B_2	B_3
	A_1	3	1	6
	A_2	2	5	3
	A_3	0	2	3

(with *B* heading spanning B_1, B_2, B_3)

d

A		B_1	B_2	B_3
	A_1	-6	-2	1
	A_2	3	-1	-2
	A_3	-4	3	1

(with *B* heading spanning B_1, B_2, B_3)

e

A		B_1	B_2	B_3	B_4
	A_1	-2	2	-3	-1
	A_2	-1	2	1	-1
	A_3	-2	-3	-4	-3

(with *B* heading spanning B_1, B_2, B_3, B_4)

f

A		B_1	B_2	B_3
	A_1	3	4	2
	A_2	-1	2	1
	A_3	1	2	4
	A_4	2	1	-3

(with *B* heading spanning B_1, B_2, B_3)

2 Xavier and Yolanda play a game in which Xavier chooses a number from the set $\{3, 5, 9\}$ and Yolanda a number from the set $\{2, 6, 7\}$. The difference between the chosen numbers, ignoring minus signs, is d. If $d > 2$ Yolanda pays Xavier £d, and otherwise Xavier pays Yolanda £$2d$.

   a i Construct Xavier's payoff matrix for this game.
   ii Use dominance to simplify the matrix as far as possible. Explain how you know the game has no stable solution.

   b Modify the game by changing one of the numbers in Yolanda's set, so that the revised game has a stable solution.

Suppose you play a game where the probability of winning is $\frac{1}{4}$. If you win, you get £10. If you lose you pay £5.

	Win	Lose
Payoff	£10	–£5
Probability	$\frac{1}{4}$	$\frac{3}{4}$

If you play the game four times, you expect to win once and lose three times.

Your expected total payoff is £10 + 3 × (–£5) = –£5
Dividing by 4, your expected average payoff per game is

$$\frac{1}{4} \times £10 + \frac{3}{4} \times (-£5) = -£5 \div 4 = -£1.25$$

This is the sum of each payoff times its probability.

> If payoffs $x_1, x_2, \ldots, x_n$ occur with probabilities $p_1, p_2, \ldots, p_n$, the **expectation** or **expected (mean) payoff** $E(x)$ is given by
>
> $$E(x) = x_1 p_1 + x_2 p_2 + \cdots + x_n p_n = \sum_{i=1}^{n} x_i p_i$$

Expectation is sometimes called expected mean.

Now consider this two-person zero-sum game.
It has no stable solution.
The play safe strategies are A_2 for A and B_1 for B, but if B knows A will play A_2, he can gain by using B_2 instead.
A knows this and counters by sometimes using A_1.
This is a **mixed strategy** game.

		B	
		B_1	B_2
A	A_1	–3	5
	A_2	3	–1

A plays A_1 randomly for a proportion p of plays, and A_2 for the remaining $(1 - p)$. She needs to decide the best value of p.

Let the value of the game be v.

> The value of a game is the payoff to the row player if both players use their best strategy.

If B plays B_1, the possible payoffs for A are shown in Table 1:

The expected payoff is $(-3) \times p + 3 \times (1 - p) = 3 - 6p$
The value of the game cannot be more than this, so $v \leqslant 3 - 6p$

	A_1	A_2
Payoff	–3	3
Probability	p	$(1 - p)$

Table 1

If B plays B_2, the possible payoffs for A are shown in Table 2:

The expected payoff is $5 \times p + (-1) \times (1 - p) = 6p - 1$
The value of the game cannot be more than this, so $v \leqslant 6p - 1$

	A_1	A_2
Payoff	5	–1
Probability	p	$(1 - p)$

Table 2

Player A needs to maximise v subject to
$v \leqslant 3 - 6p$ and $v \leqslant 6p - 1$.

Plot lines $v = 3 - 6p$ and $v = 6p - 1$:

For a given value of p, the value of v occurs
on the lower of the two lines.

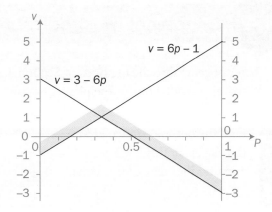

Maximum v occurs where the two lines intersect,
that is where

$$3 - 6p = 6p - 1 \quad \text{giving} \quad p = \frac{1}{3}$$

A's best strategy is to play A_1 and A_2 randomly
with probabilities $\frac{1}{3}$ and $\frac{2}{3}$.

The value of the game is $v = 3 - 6 \times \frac{1}{3} = 1$

You can find B's best strategy in the same way.
Suppose B plays B_1 and B_2 with probabilities q and $(1 - q)$.

If A plays A_1, B's expected payoff is

$$3 \times q + (-5) \times (1 - q)) = 8q - 5$$

If A plays A_2, B's expected payoff is

$$(-3) \times q + 1 \times (1 - q)) = 1 - 4q$$

The optimal strategy occurs when these are equal,

i.e. when $8q - 5 = 1 - 4q$ $\quad$ giving $\quad q = \frac{1}{2}$

B's best strategy is to play B_1 and B_2 half the time each, at random.

The value of the game to B is $8 \times \frac{1}{2} - 5 = -1$

> This is a linear programming situation. This is explored more fully in Section 4.5.

> The table shows A's payoffs. B's payoffs are the negative of these.

> This is $-v$, as you would expect.

D2

Exercise 4.3

1 For the following games decide if a mixed strategy is needed.
   Find the optimal strategy (pure or mixed) for both players
   and the value of the game.

a

		\multicolumn{2}{c}{B}
		B_1	B_2
A	A_1	-2	4
	A_2	3	2

b

		\multicolumn{2}{c}{B}
		B_1	B_2
A	A_1	4	2
	A_2	-1	3

c

		\multicolumn{2}{c}{B}
		B_1	B_2
A	A_1	5	-3
	A_2	3	2

d

		\multicolumn{2}{c}{B}
		B_1	B_2
A	A_1	1	4
	A_2	4	3

You can generalise the methods of Section 4.3 to larger payoff matrices provided one of the players has just two strategies.

Find the optimal strategies for the players in this 2 × 3 game, and find the value of the game.

		B		
		B_1	B_2	B_3
A	A_1	4	2	-1
	A_2	-3	-2	5

The game has no stable solution, so A plays A_1 and A_2 with probabilities p and $(1 - p)$.

If B plays B_1, A's expectation $= 4p - 3(1 - p) = 7p - 3$
If B plays B_2, A's expectation $= 2p - 2(1 - p) = 4p - 2$
If B plays B_3, A's expectation $= -p + 5(1 - p) = 5 - 6p$

If the value of the game is v, you need to maximise v subject to
$v \leqslant 7p - 3$,
$v \leqslant 4p - 2$,
$v \leqslant 5 - 6p$

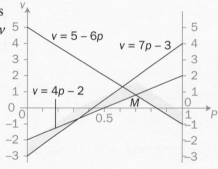

Maximum v is at vertex M on the graph,
where $v = 4p - 2$ intersects with $v = 5 - 6p$
$\quad 4p - 2 = 5 - 6p \quad$ so $\quad p = 0.7$

So A plays A_1 and A_2 randomly with probabilities 0.7 and 0.3.
The value of the game is $v = 4 \times 0.7 - 2 = 0.8$

Line $v = 7p - 3$ passes above M, so B should not play B_1 because A's payoff would be greater than 0.8

B plays B_2 and B_3 with probabilities q and $(1 - q)$.

If A plays A_1, B's expectation is $-2q + (1 - q) = 1 - 3q$
If A plays A_2, B's expectation is $2q - 5(1 - q) = 7q - 5$

The optimal strategy occurs where these lines intersect.
When $\quad 1 - 3q = 7q - 5, \quad q = 0.6$
So B plays B_2 and B_3 with probabilities 0.6 and 0.4
The value of the game to B is $1 - 3 \times 0.6 = -0.8$, as expected.

In general in a 2 × *n* game the column player will only use two of the *n* available strategies, as the others will give the row player a greater payoff (but see Ex 4.4 qu **5** for a special case).

The table shows A's payoffs, and B's payoffs are the negative of these.

In an $n \times 2$ game you analyse the situation from the point of view of the column player.

EXAMPLE 2

Find the optimal strategies for the players in this 3×2 game, and find the value of the game.

		B	
		B_1	B_2
A	A_1	-4	1
	A_2	0	-2
	A_3	1	-4

The game has no stable solution, so B plays B_1 and B_2 with probabilities q and $(1-q)$.

If A plays A_1, B's expectation $= 4q - (1-q) = 5q - 1$
If A plays A_2, B's expectation $= 0 \times q + 2(1-q) = 2 - 2q$
If A plays A_3, B's expectation $= -q + 4(1-q) = 4 - 5q$

The table shows A's payoffs. B's payoffs are the negative of these.

The value of the game (to A) is v, so the value to B is $(-v)$. You need to maximise $(-v)$ subject to
$(-v) \leqslant 5q - 1$, $(-v) \leqslant 2 - 2q$, $(-v) \leqslant 4 - 5q$
This is equivalent to minimising v subject to
$v \geqslant 1 - 5q$,
$v \geqslant 2q - 2$,
$v \geqslant 5q - 4$

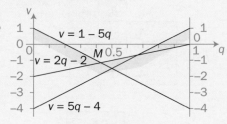

Minimum v is at vertex M on the graph, where $v = 2q - 2$ intersects with $v = 1 - 5q$

When $2q - 2 = 1 - 5q$, $q = \frac{3}{7}$

So B plays B_1 and B_2 with probabilities $\frac{3}{7}$ and $\frac{4}{7}$.

The value of the game is $v = 2 \times \frac{3}{7} - 2 = -1\frac{1}{7}$

The value of the game to B is $1\frac{1}{7}$.

Line $v = 5q - 4$ passes below M, so A should not play A_3 because B's payoff would be greater than $1\frac{1}{7}$.

A plays A_1 and A_2 with probabilities p and $(1-p)$.

If B plays B_1, A's expectation is $-4p + 0 \times (1-p) = -4p$
If B plays B_2, A's expectation is $p - 2(1-p) = 3p - 2$

The optimal strategy occurs where these lines intersect.
When $-4p = 3p - 2$, $p = \frac{2}{7}$

In general in an $n \times 2$ game the row player will only use two of the n available strategies, as the others will give the column player a greater payoff.

D2

Exercise 4.4

1 For each of these 2 × 3 games, find the optimal mixed strategy for each player and the value of the game.

a

		B		
		B_1	B_2	B_3
A	A_1	4	2	-2
	A_2	-3	-2	3

b

		B		
		B_1	B_2	B_3
A	A_1	7	6	9
	A_2	5	8	4

c

		B		
		B_1	B_2	B_3
A	A_1	5	2	1
	A_2	-1	3	8

d

		B		
		B_1	B_2	B_3
A	A_1	-2	4	1
	A_2	2	-1	0

2 For each of these 3 × 2 games, find the optimal mixed strategy for each player and the value of the game.

a

		B	
		B_1	B_2
	A_1	-1	5
A	A_2	4	-3
	A_3	0	1

b

		B	
		B_1	B_2
	A_1	-6	2
A	A_2	-5	1
	A_3	1	-2

c

		B	
		B_1	B_2
	A_1	9	7
A	A_2	3	10
	A_3	6	8

d

		B	
		B_1	B_2
	A_1	-5	2
A	A_2	1	-4
	A_3	-3	-2

3 The table shows a game in which player A has three strategies A_1, A_2 and A_3, while player B has four strategies B_1, B_2, B_3 and B_4.

a Show that this game has no saddle point.

b Use dominance arguments to reduce the matrix as far as possible.

c Find the optimal mixed strategy for player A and the value of the game.

d Find the optimal mixed strategy for player B.

		B			
		B_1	B_2	B_3	B_4
	A_1	-1	3	7	-1
A	A_2	1	5	4	2
	A_3	4	2	3	3

D2

4 The table shows a game between players X and Y, each with four possible strategies.

		Y			
		I	II	III	IV
X	I	1	4	3	2
	II	-1	3	0	3
	III	4	2	5	1
	IV	-2	6	-3	1

a Show that this game has no saddle point.

b Use dominance arguments to reduce the matrix as far as possible.

c Find the optimal mixed strategy for each player and the value of the game.

5 The table shows a game between players A and B. A has two strategies: A_1 and A_2. B has three strategies: B_1, B_2 and B_3.

		B		
		B_1	B_2	B_3
A	A_1	2	5	6
	A_2	5	2	1

a Use a graph to show that player A's optimal strategy is to play her two strategies at random with equal probability.

b Find the value of the game.

c Explain, with reference to your graph from part **a**, why in this game player B could reasonably make use of all three available strategies.

d Let B play his strategies with probabilities q_1, q_2 and q_3 respectively.

   i Making use of the known value of the game, write down three equations connecting q_1, q_2 and q_3.

   ii Putting $q_1 = q$, show that $q_2 = 2\frac{1}{2} - 4q$ and find q_3 in terms of q.

   iii Show that $\frac{1}{2} \leqslant q \leqslant \frac{5}{8}$ and find the range of possible values of q_2 and q_3.

D2

To solve a game involving a mix of three or more strategies you must write it in linear programming terms. You could then use the simplex algorithm to obtain a solution.

The simplex algorithm limits the problem in two ways:

- It must be a maximisation problem.
- No decision variable can be negative.

These restrictions mean that you must take a different approach depending on whether you are analysing the game from the point of view of the row player or the column player.

For the row player the problem is a maximisation one – you are trying to maximise the value, v, of the game. However, as v appears as one of the decision variables, you may need to adjust the situation so that v is positive.

> In the exam you will be required to formulate the problem in linear programming terms, but you will not be asked to solve it.

EXAMPLE 1

Express the problem of finding the optimal mixed strategy for player A as a linear programming formulation. Write the problem in the form of equations with slack variables.

		B		
		B_1	B_2	B_3
	A_1	2	-2	-3
A	A_2	-3	-4	4
	A_3	1	2	3

Some of A's payoffs are negative, so it is possible that the value of the game is negative.

To overcome this difficulty, add 4 to each payoff:

		B		
		B_1	B_2	B_3
	A_1	6	2	1
A	A_2	1	0	8
	A_3	5	6	7

Let the value of this game be v, and let A play strategies A_1, A_2 and A_3 with probabilities p_1, p_2 and p_3.

> If the value of the original game is V, the value of the modified game is $v = V + 4$. Having solved the problem, you reduce the value found by 4 for the final answer.

The objective function, P, is the value, v, of the game. The value is limited by A's expected payoffs when B plays each of his three strategies:

$$\text{Maximise} \quad P = v$$
$$\text{subject to} \quad v \leqslant 6p_1 + p_2 + 5p_3$$
$$v \leqslant 2p_1 + 6p_3$$
$$v \leqslant p_1 + 8p_2 + 7p_3$$
$$p_1 + p_2 + p_3 \leqslant 1$$
$$v, p_1, p_2, p_3 \geqslant 0$$

> $p_1 + p_2 + p_3 = 1$, but it is usual to write an inequality so that it appears as an equation with a slack variable in the standard simplex layout.

EXAMPLE 1 (CONT.)

Using slack variables r, s, t and u, this becomes

Maximise $P = v$
subject to $v - 6p_1 - p_2 - 5p_3 + r = 0$
$v - 2p_1 - 6p_3 + s = 0$
$v - p_1 - 8p_2 - 7p_3 + t = 0$
$p_1 + p_2 + p_3 + u = 1$
$v, p_1, p_2, p_3 \geqslant 0$

This formulation can be refined to reduce the number of decision variables from four to three.

$p_1 + p_2 + p_3 = 1$,   so $p_3 = 1 - p_1 - p_2$

Substitute into the first constraint:
$v \leqslant 6p_1 + p_2 + 5(1 - p_1 - p_2)$
$v - p_1 + 4p_2 \leqslant 5$

Continuing in this way, the problem becomes:

Maximise $P = v$
subject to $v - p_1 + 4p_2 \leqslant 5$
$v + 4p_1 + 6p_2 \leqslant 6$
$v + 6p_1 - p_2 \leqslant 7$
$v, p_1, p_2 \geqslant 0$

Using slack variables r, s and t, this becomes

Maximise $P = v$
subject to $v - p_1 + 4p_2 + r = 5$
$v + 4p_1 + 6p_2 + s = 6$
$v + 6p_1 - p_2 + t = 7$
$v, p_1, p_2 \geqslant 0$

The formulation obtained already is sufficient in an examination unless the question asks you to go further.

Check that you can derive these inequalities.

D2

For the column player you have a minimisation problem – to minimise v so that the row player gains as little as possible.

You convert to a maximisation problem by making $\frac{1}{v}$ the objective function. Maximising $\frac{1}{v}$ is equivalent to minimising v.

You could just rewrite the game so that the column player becomes the row player, changing the signs of all the payoffs, but changing the objective function is a more satisfactory approach.

EXAMPLE 2

Express the problem of finding the optimal mixed strategy for player B as a linear programming formulation. Write the problem in the form of equations with slack variables.

		B		
		B_1	B_2	B_3
A	A_1	2	-2	-3
	A_2	-3	-4	4
	A_3	1	2	3

You need to add 4 to each payoff so v and therefore $\frac{1}{v}$ cannot be negative.

Let B play strategies B_1, B_2 and B_3 with probabilities q_1, q_2 and q_3.

		B		
		B_1	B_2	B_3
A	A_1	6	2	1
	A_2	1	0	8
	A_3	5	6	7

$$q_1 + q_2 + q_3 = 1 \quad \text{so} \quad \frac{q_1}{v} + \frac{q_2}{v} + \frac{q_3}{v} = \frac{1}{v}$$

Define new variables x_1, x_2 and x_3 as $x_1 = \frac{q_1}{v}$, $x_2 = \frac{q_2}{v}$, $x_3 = \frac{q_3}{v}$:

The objective function $P = \frac{1}{v} = x_1 + x_2 + x_3$

The value to B is $(-v)$. This is limited by B's expected payoffs when A plays each of her three strategies:

$$(-v) \leqslant -6q_1 - 2q_2 - q_3$$

Divide by $(-v)$:

$$1 \geqslant 6\frac{q_1}{v} + 2\frac{q_2}{v} + \frac{q_3}{v} \quad \text{so} \quad 6x_1 + 2x_2 + x_3 \leqslant 1$$

Continuing in this way, the complete formulation is

Maximise $\quad P = x_1 + x_2 + x_3$

subject to $\quad 6x_1 + 2x_2 + x_3 \leqslant 1$

$\qquad\qquad\qquad x_1 + 8x_3 \leqslant 1$

$\qquad\qquad 5x_1 + 6x_2 + 7x_3 \leqslant 1$

$\qquad\qquad\qquad x_1, x_2, x_3 \geqslant 0$

Using slack variables r, s and t this becomes

Maximise $\quad P = x_1 + x_2 + x_3$

subject to $\quad 6x_1 + 2x_2 + x_3 + r = 1$

$\qquad\qquad\qquad x_1 + 8x_3 + s = 1$

$\qquad\qquad 5x_1 + 6x_2 + 7x_3 + t = 1$

$\qquad\qquad\qquad x_1, x_2, x_3, \geqslant 0$

Remember to reverse the inequality sign when you divide by $-v$.

D2

Exercise 4.5

1 Formulate each of the following games as a linear programming problem for player A.

Express your constraints as equations with slack variables.

a

		B	
		B_1	B_2
	A_1	5	1
A	A_2	2	4
	A_3	3	2

b

		B	
		B_1	B_2
	A_1	–3	1
A	A_2	–1	–4
	A_3	0	–5

c

		B		
		B_1	B_2	B_3
	A_1	2	1	3
A	A_2	1	4	4
	A_3	5	2	1

d

		B		
		B_1	B_2	B_3
	A_1	–8	–2	1
A	A_2	–3	–1	–4
	A_3	0	–3	–2

2 Formulate each of the following games as a linear programming problem for player B.

Write your constraints as inequalities and define your variables clearly.

a

		B		
		B_1	B_2	B_3
	A_1	5	3	1
A	A_2	1	2	6

b

		B		
		B_1	B_2	B_3
	A_1	–4	–2	–1
A	A_2	–1	3	–5

c

		B		
		B_1	B_2	B_3
	A_1	3	3	2
A	A_2	1	5	4
	A_3	4	2	3

d

		B		
		B_1	B_2	B_3
	A_1	–2	–1	1
A	A_2	–1	–3	–4
	A_3	0	–5	1

3 The table shows a game with value V.

		B		
		B_1	B_2	B_3
	A_1	–1	0	–2
A	A_2	1	–2	–1
	A_3	0	–1	1

a Player A plays A_1, A_2 and A_3 with probabilities p_1, p_2 and p_3.

  i Express the game as a linear programming problem for player A using just two of the probabilities.

  ii Set up and solve the corresponding simplex tableau.
State A's optimal strategy and the value, V, of the original game.

b Player B plays B_1, B_2 and B_3 with probabilities q_1, q_2 and q_3.

  i Express the game as a linear programming problem for player B.

  ii Set up and solve the corresponding simplex tableau.
Verify the value of the game and state B's optimal strategy.

1 a Explain what is meant by a zero-sum game.

   b A two-person zero-sum game is represented by
      this payoff matrix for player A.

      i   Find the play safe strategy for each player.
      ii   Verify that the game has a saddle point.
           Explain the implications of this.
      iii State the value of the game.

		B		
		I	II	III
	I	1	0	-2
A	II	-1	4	0
	III	2	3	5

   c Write down the payoff matrix for player B.

2 The table shows the payoff matrix for player A in a
   two-person zero-sum game.

		B	
		B_1	B_2
A	A_1	-1	3
	A_2	4	2

   a Verify that there is no stable solution to this game.

   b Find the optimal strategy for player A and the value
      of the game to her.

   c Find the optimal strategy for player B.

3 The table shows the payoff matrix for player A in a
   two-person zero-sum game.

		B		
		I	II	III
	I	5	3	9
A	II	4	7	6
	III	2	4	5

   a Use dominance arguments to reduce the matrix as
      far as possible.

   b Show that the game has no saddle point.

   c Find the optimal strategy for each player and the
      value of the game.

4 A two-person zero-sum game is represented by this payoff matrix for player *A*.

a Use graphical methods to find *A*'s optimal strategy and the value of the game.

b Explain why *B* will not use one of the three available strategies. Hence find *B*'s optimal strategy.

		B		
		B_1	B_2	B_3
A	A_1	2	3	1
	A_2	3	2	5

5 A two-person zero-sum game is represented by this payoff matrix for player *A*.

a Verify that there is no stable solution for this game.

b Find graphically the optimal strategy for player *B*.

c Explain why player *A* should not use strategy I. Find the optimal mix of the remaining two strategies.

d State the value of the game.

		B	
		I	II
	I	–2	4
A	II	–1	2
	III	4	1

6 A two-person zero-sum game is represented by this payoff matrix for player *A*.

a Find the optimal strategy for player *B*.

b Formulate the game as a linear programming problem for player *A*. Write your constraints as equations with slack variables.

		B	
		B_1	B_2
	A_1	2	–4
A	A_2	–3	1
	A_3	–1	–3

D2

7 Mark and Nasreen play a zero-sum game which is represented by this payoff matrix for Mark.

		N			
		I	II	III	IV
	I	–2	1	3	–1
	II	2	–1	1	2
M	III	–3	2	1	–2
	IV	–3	0	2	–1

a Use dominance arguments to reduce the game to a 3 × 3 matrix.

b Using your reduced matrix, formulate the game as a linear programming problem for Nasreen. Write your constraints as inequalities.

4

Exit ⟹

Summary

Refer to

- A play safe strategy gives the best guaranteed outcome in a game.
  - For the row player, maximise the row minima (maximin).
  - For the column player, minimise the column maxima (minimax).
- The value of a game is the payoff to the row player if both players use their best strategy.
- A game has a stable solution (a saddle point) if neither player can gain by changing from their play safe strategy. In this case
  max of row minima = min of column maxima = value of the game.   4.1
- A row or column can be eliminated if it is dominated by another.   4.2
- In an optimal mixed strategy, strategies are played at random with probabilities chosen to give the maximum expected payoff.   4.3
- For a $2 \times n$ or $n \times 2$ game:
  - Use graphical methods for the player with two strategies.
  - The other player rejects all but two strategies. Analyse graphically.   4.4
- Larger matrices need linear programming formulation. If there are negative payoffs, add a constant to all payoffs.   4.5

Links

Game theory has many applications in real life, including in politics, social science and economics.

In finance, when fund managers make an investment decision they choose between different investment strategies. Each different strategy achieves a different return, depending on the future state of the market. This situation could be analysed as a two-person game between the fund manager and the market.

D2

1 A company makes three sizes of lamps, small, medium and large. The company is trying to determine how many of each size to make in a day, in order to maximise its profit. As part of the process the lamps need to be sanded, painted, dried and polished. A single machine carries out these tasks and is available 24 hours per day. A small lamp requires one hour on this machine, a medium lamp 2 hours and a large lamp 4 hours.

Let x = number of small lamps made per day
y = number of medium lamps made per day
z = number of large lamps made per day
where $x \geqslant 0$, $y \geqslant 0$ and $z \geqslant 0$.

a Write the information about this machine as a constraint.

b i Rewrite your constraint from part a using a slack variable s.
ii Explain what s means in practical terms.

Another constraint and the objective function give the following simplex tableau. The profit P is stated in euros.

Basic variable	x	y	z	r	s	Value
r	3	5	6	1	0	50
s	1	2	4	0	1	24
P	−1	−3	−4	0	0	0

c Write down the profit on each small lamp.

d Use the simplex algorithm to solve this linear programming problem.

e Explain why the solution to part d is not practical.

f Find a practical solution which gives a profit of 30 euros. Verify that it is feasible.

[(c) Edexcel Limited 2003]

2 T42 Co. Ltd produces three different blends of tea, Morning, Afternoon and Evening. The teas must be processed, blended and then packed for distribution. The table shows the time taken, in hours, for each stage of the production of a tonne of tea. It also shows the profit, in hundreds of pounds, on each tonne.

	Processing	Blending	Packing	Profit (£100)
Morning blend	3	1	2	4
Afternoon blend	2	3	4	5
Evening blend	4	2	3	3

The total times available each week for processing, blending and packing are 35, 20 and 24 hours respectively. T42 Co. Ltd wishes to maximise the weekly profit.

Let x, y and z be the number of tonnes of Morning, Afternoon and Evening blend produced each week.

a   Formulate the above situation as a linear programming problem, listing clearly the objective function, and the constraints as inequalities.

An initial simplex tableau for this situation is

Basic variable	x	y	z	r	s	t	Value
r	3	2	4	1	0	0	35
s	1	3	2	0	1	0	20
t	2	4	3	0	0	1	24
P	−4	−5	−3	0	0	0	0

b   Solve this linear programming problem using the simplex algorithm. Take the most negative number in the profit row to indicate the pivot column at each stage.

T42 Co. Ltd wishes to increase its profit further and is prepared to increase the time available for processing or blending or packing or any two of these three.

c   Use your answer to part b to advise the company as to which stage(s) it should increase the time available.

[(c) Edexcel Limited 2002]

3 While solving a maximising linear programming problem, this tableau was obtained.

Basic variable	x	y	z	r	s	t	Value
r	0	0	$1\frac{2}{3}$	1	0	$-\frac{1}{6}$	$\frac{2}{3}$
y	0	1	$3\frac{1}{3}$	0	1	$-\frac{1}{3}$	$\frac{1}{3}$
x	1	0	-3	0	-1	$\frac{1}{2}$	1
P	0	0	1	0	1	1	11

a Explain why this is an optimal tableau.

b Write down the optimal solution of this problem, stating the value of every variable.

c Write down the profit equation from the tableau. Use it to explain why changing the value of any of the non-basic variables will decrease the value of P.

[(c) Edexcel Limited 2002]

4 This tableau is the initial tableau for a linear programming problem in x, y and z. The objective is to maximise the profit, P.

Basic variable	x	y	z	r	s	t	Value
r	12	4	5	1	0	0	246
s	9	6	3	0	1	0	153
t	5	2	-2	0	0	1	171
P	-2	-4	-3	0	0	0	0

Using the information in the tableau, write down

a the objective function

b the three constraints as inequalities with integer coefficients.

Taking the most negative number in the profit row to indicate the pivot column at each stage

c solve this linear programming problem. Make your method clear by stating the row operations you use.

d State the final values of the objective function and each variable.

One of the constraints is not at capacity.

e Explain how it can be identified.

[(c) Edexcel Limited 2007]

5 Freezy Co. has three factories *A*, *B* and *C*. It supplies freezers to three shops *D*, *E* and *F*. The table shows the transportation cost in pounds of moving one freezer from each factory to each outlet. It also shows the number of freezers available for delivery at each factory and the number of freezers required at each shop. The total number of freezers required is equal to the total number of freezers available.

	D	E	F	Available
A	21	24	16	24
B	18	23	17	32
C	15	19	25	14
Required	20	30	20	

a   Use the north-west corner rule to find an initial solution.

b   Obtain improvement indices for each unused route.

c   Use the stepping-stone method **once** to obtain a better solution and state its cost.

[(c) Edexcel Limited 2005]

6 a   Describe a practical problem that could be solved using the transportation algorithm.

A problem is to be solved using the transportation algorithm. The costs are shown in the table. The supply is from *A*, *B* and *C* and the demand is at *d* and *e*.

	d	e	Supply
A	5	3	45
B	4	6	35
C	2	4	40
Demand	50	60	

b   Explain why it is necessary to add a third demand *f*.

c   Use the north-west corner rule to obtain a possible pattern of distribution and find its cost.

d   Calculate shadow costs and improvement indices for this pattern.

e   Use the stepping-stone method **once** to obtain an improved solution and its cost.

[(c) Edexcel Limited 2004]

7 a Explain briefly the circumstances under which a degenerate feasible solution may occur to a transportation problem.

b Explain why a dummy location may be needed when solving a transportation problem.

The table shows the cost of transporting one unit of stock from each of three supply points A, B and C to each of two demand points 1 and 2. It also shows the stock held at each supply point and the stock required at each demand point.

	1	2	Supply
A	62	47	15
B	61	48	12
C	68	58	7
Demand	16	11	

c Find an initial feasible solution using the north-west corner method.

d Use the stepping-stone method to obtain an optimal solution and state its cost. You should make your method clear by stating shadow costs, improvement indices, stepping-stone route, and the entering and exiting squares at each stage.

[(c) Edexcel Limited 2006]

D2

8 Three depots, F, G and H, supply petrol to three service stations S, T and U. The table gives the cost, in pounds, of transporting 1000 litres of petrol from each depot to each service station.

	S	T	U
F	23	31	46
G	35	38	51
H	41	50	63

F, G and H have stocks of 540 000, 789 000 and 673 000 litres respectively. S, T and U require 257 000, 348 000 and 412 000 litres respectively.

The total cost of transporting the petrol is to be minimised.

Formulate this problem as a linear programming problem. Make clear your decision variables, objective function and constraints.

[(c) Edexcel Limited 2006]

9 Talkalot College holds an induction meeting for new students. The meeting consists of four talks: I (Welcome), II (Options and Facilities), III (Study Tips) and IV (Planning for Success). The four department heads, Clive, Julie, Nicky and Steve, deliver one of these talks each. The talks are delivered consecutively and there are no breaks between talks. The meeting starts at 10 a.m. and ends when all four talks have been delivered. The time, in minutes, each department head takes to deliver each talk is given in the table.

	Talk I	Talk II	Talk III	Talk IV
Clive	12	34	28	16
Julie	13	32	36	12
Nicky	15	32	32	14
Steve	11	33	36	10

a Use the Hungarian algorithm to find the earliest time that the meeting could end. You must make your method clear and show

   i the state of the table after each stage in the algorithm

   ii the final allocation.

b Modify the table so it could be used to find the latest time that the meeting could end. (You do not have to find this latest time.) [(c) Edexcel Limited 2003]

10 To raise money for charity it is decided to hold a Teddy Bear Making competition. Teams of four compete against each other to make 20 teddy bears as quickly as possible.

There are four stages: first cutting, then stitching, then filling and finally, dressing. Each team member can only work on one stage during the competition. As soon as a stage is completed on each teddy bear the work is passed immediately to the next team member.

The table shows the time, in seconds, taken to complete each stage of the work on one teddy bear by the members A, B, C and D of one of the teams.

	Cutting	Stitching	Filling	Dressing
A	66	101	85	36
B	66	98	74	38
C	63	97	71	34
D	67	102	78	35

a Use the Hungarian algorithm, reducing rows first, to obtain an allocation that minimises the time taken by this team to produce one teddy bear. You must make your method clear and show the table after each iteration.

b State the minimum time it will take this team to produce one teddy bear.

[(c) Edexcel Limited 2007]

11 A theme park has four sites, *A*, *B*, *C* and *D*, on which to put kiosks. Each kiosk will sell a different type of refreshment. The income from each kiosk depends on what it sells and where it is located. The table shows the expected daily income, in pounds, from each kiosk at each site.

	Hot dogs and beefburgers (*H*)	Ice cream (*I*)	Popcorn, candyfloss and drinks (*P*)	Snacks and hot drinks (*S*)
Site *A*	267	272	276	261
Site *B*	264	271	278	263
Site *C*	267	273	275	263
Site *D*	261	269	274	257

Reducing rows first, use the Hungarian algorithm to determine a site for each kiosk in order to maximise the total income. State the site for each kiosk and the total expected income. You must make your method clear and show the table after each stage.

[(c) Edexcel Limited 2006]

12 a Explain briefly what is meant by a zero-sum game.

A two person zero-sum game is represented by the following pay-off matrix for player *A*.

	I	II	III
I	5	2	3
II	3	5	4

b Verify that there is no stable solution to this game.

c Find the best strategy for player *A* and the value of the game to her.

d Formulate the game as a linear programming problem for player *B*. Write the constraints as inequalities and define your variables clearly.

[(c) Edexcel Limited 2005]

D2

13 A two-person zero-sum game is represented by the following pay-off matrix for player A.

		B			
		I	II	III	IV
	I	-4	-5	-2	4
A	II	-1	1	-1	2
	III	0	5	-2	-4
	IV	-1	3	-1	1

a Determine the play-safe strategy for each player.

b Verify that there is a stable solution and determine the saddle points.

c State the value of the game to B.   [(c) Edexcel Limited 2002]

14 Denis (D) and Hilary (H) play a two-person zero-sum game represented by the following pay-off matrix for Denis.

	H plays 1	H plays 2	H plays 3
D plays 1	2	-1	3
D plays 2	-3	4	-4

a Show that there is no stable solution to this game.

b Find the best strategy for Denis and the value of the game to him.   [(c) Edexcel Limited 2007]

15 A two-person zero-sum game is represented by the following payoff matrix for player A. Find the best strategy for each player and the value of the game.

		B	
		I	II
A	I	4	-2
	II	-5	6

[(c) Edexcel Limited 2007]

5

Travelling salesman problem

This chapter will enable you to
- identify a Hamiltonian cycle
- convert a given network into the corresponding complete network of shortest distances
- find an upper bound for the optimal tour using a minimum spanning tree
- improve an upper bound using short cuts
- find an upper bound using the nearest neighbour algorithm
- find a lower bound for the optimal tour by using a minimum spanning tree for a reduced network.

Introduction

In the route inspection problem the aim is to travel along every edge of the network.

Refer to D1 for revision of networks and minimum spanning trees.

In the travelling salesman problem the aim is to visit every vertex. The problem gets its name from the idea of a sales representative attending appointments at a number of different locations before returning to base.

This chapter uses Kruskal's algorithm and Prim's algorithm in relation to a minimum spanning tree (minimum connector).

D2

Recall that

- a path is a continuous route within a graph which does not repeat any vertices
- a cycle (or circuit) is a closed path (it finishes where it started).

A Hamiltonian cycle (tour) is a closed path which visits every vertex of the graph.

It is named after the Irish mathematician Sir William Rowan Hamilton (1805–1865).

A Hamiltonian graph has at least one Hamiltonian cycle.

In this diagram, *ACEGBDFA* is a Hamiltonian cycle.

Any vertex could be taken as the start of the cycle. For example, *GBDFACEG* is the same cycle.

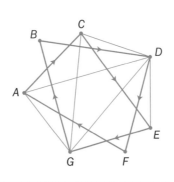

Some graphs do not possess a Hamiltonian cycle.

Any complete tour of this graph would need to visit *C* twice. The graph is not Hamiltonian.

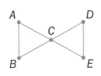

There is as yet no general test for deciding whether a given graph is Hamiltonian.

The graph shown is not Hamiltonian. Show that by adding one extra edge you can make it Hamiltonian.

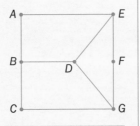

Add the edge *DF*:

There is now a Hamiltonian cycle *ABCGDFEA*.

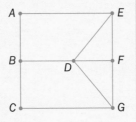

There are a number of possible answers to this. You might like to try some other possibilities.

Exercise 5.1

1 For each of the following graphs, state whether or not there is
   a Hamiltonian cycle. If there is, give an example.

a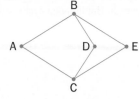

b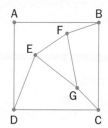

c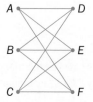

d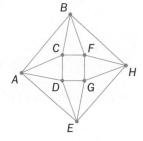

2 Write down a Hamiltonian cycle for the graph shown.

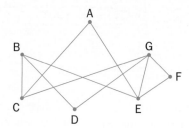

3 The complete graph K_n is Hamiltonian for all $n > 2$.

   a Find the number of different Hamiltonian cycles there are in

   i K_3 ii K_4 iii K_5

   b Find a formula for the number of different Hamiltonian cycles in K_n.

Refer to **D1** for revision of
complete graphs.

The **classical TSP** is to visit every vertex of a network *once only*, and return to the start in the least possible distance.

> You want to find a minimum weight Hamiltonian cycle.

The **practical TSP** may not fit the classical pattern in two ways.

Firstly, there may not be a Hamiltonian cycle:

There is no Hamiltonian cycle in this network.

By inspection the best route is $ABDCDBEBA = 60$

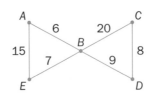

> A network satisfies the triangle inequality if for every triangle in the network the sum of the weights of any two sides is at least as big as the weight of the third side.

Secondly, the network may not satisfy the triangle inequality:

This network has a Hamiltonian cycle $ABCA = 36$, but the best practical solution to the problem is $ACBCA = 32$

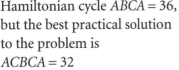

To solve a practical TSP with n vertices you convert it to an equivalent classical TSP on the complete graph K_n. The weight on each edge of K_n corresponds to the shortest distance between those vertices on the original graph. Solving the classical problem leads to the solution to the practical problem.

> In complex cases you could find the shortest distances using Dijkstra's algorithm, but in simple cases you can do this by inspection.

Solving a practical TSP
- Create a complete network of shortest distances.
- Find a minimum weight Hamiltonian cycle for the complete network.
- Interpret the solution in the practical situation.

D2

EXAMPLE 1

Convert this network to a complete network of shortest distances. By solving the classical TSP for the complete network, find a solution to the practical problem.

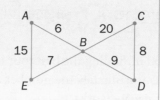

There are 5 vertices, so draw K_5. The weights are shortest distances, so for example

   weight $AE = 13$

because $ABE = 13$ is the shortest route on the original network.

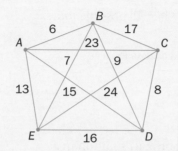

Similarly
   weight $CE = 24$
because $CDE = 24$

A minimum Hamiltonian tour can be found by inspection.
   $ABDCEA = 60$

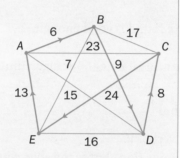

This is not the only Hamiltonian tour of length 60 in this network. Try finding others. You should discover that they all lead to the same practical solution.

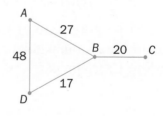

The solution to the practical problem is $ABDCDBEBA = 60$

D2

Exercise 5.2

1 The diagram shows roads connecting four towns, with distances in km. A lorry needs to visit each town, starting and finishing at A.

a Draw a complete network of shortest distances.

b List the three distinct Hamiltonian tours of your complete network and show that each has a total weight of 128 km.

c Show that each of these tours corresponds to the same route between the towns. State this route.

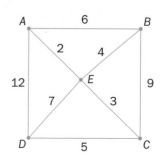

2 a List the four distinct Hamiltonian tours of this network. State the minimum total weight of such a tour.

b Draw a complete network of shortest distances.

c Show that there is a Hamiltonian tour of the complete network with total weight 27.

d Write down the route around the original network corresponding to your answer to part c.

The only known sure-fire method of finding the best tour is to check every Hamiltonian cycle. Unfortunately the number of cycles is huge for even moderate size networks.

A network with n vertices has $\frac{1}{2} \times (n-1)!$ possible tours.

For a network with 16 vertices you would need to examine
$\frac{1}{2} \times 15! = 6.54 \times 10^{11}$ different tours.

Because of this you need a way to

○ find a good (not necessarily optimal) solution in a reasonable length of time
○ decide if the solution you have found is good enough.

A simple heuristic method is the **nearest neighbour algorithm**.

> **The nearest neighbour algorithm**
>
> Step 1   Choose a starting vertex V.
> Step 2   From your current position choose the edge with minimum weight leading to an unvisited vertex. Travel to that vertex.
> Step 3   If there are unvisited vertices go to Step 2.
> Step 4   Travel back to V.
>
> If at Step 2 there are equal edges, choose at random. You can re-run the algorithm making the other choice(s) to see if they give better solutions.

You can repeat the nearest neighbour algorithm taking each vertex in turn as the starting vertex. This often produces several different tours from which you can select the best.

At the start there are $(n-1)$ edges to choose from, then $(n-2)$ second edges and so on. Each tour gets listed twice.

If a computer checked a million tours per second K_{16} would take nearly 8 days. For K_{20} this time would rise to 1928 years!

To decide if the solution is good enough you find upper and lower bounds for the optimal solution – see Sections 5.4 and 5.5.

An algorithm giving a satisfactory but perhaps not optimal solution is called a **heuristic algorithm**.

This is another example of a greedy algorithm, where you make the immediately most advantageous choice at each stage without looking ahead to its implications.

Once a tour has been found, any vertex can be its starting point. The 'starting vertex' referred to in the algorithm is the starting vertex for the process of finding the tour – it does not have to be the actual start of the salesman's journey.

EXAMPLE 1

This network shows the distances, in km, between five towns. Use the nearest neighbour algorithm, starting from each vertex in turn, to find a route for a travelling salesman, who is based at A and needs to visit every town.

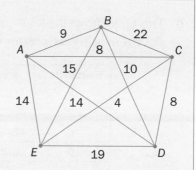

Start at vertex A:

From A, nearest vertex is C, weight $AC = 8$
From C, nearest unvisited vertex is E, $CE = 4$
From E, nearest unvisited vertex is B, $EB = 15$
From B, nearest unvisited vertex is D, $BD = 10$
No more unvisited vertices, so travel $DA = 14$

The total weight for $ACEBDA$ is
$8 + 4 + 15 + 10 + 14 = 51$

Repeat starting from the other vertices:

Start vertex	Tour(s)	Total weight
A	ACEBDA	51
B	BACEDB	50
C	CEABDC	45
D	DCEABD	45
E	ECABDE	50
	ECDBAE	45

Check that you can see how these results come about.

The three tours of length 45 km are in fact the same tour. For a salesman based at A, the best tour produced by the nearest neighbour algorithm is $ABDCEA = 45$ km.

It is in fact the optimal tour, but the nearest neighbour algorithm will not always give an optimal result.

D2

EXAMPLE 2

The diagram shows the travel times, in minutes, between six locations.

a Draw up a table showing the complete network of shortest travel times.

b Use the nearest neighbour algorithm to find a solution to the travelling salesman problem.

c Show that the solution found in part **b** is not optimal.

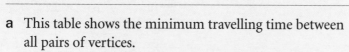

a This table shows the minimum travelling time between all pairs of vertices.

Drawing K_6 gives a very cluttered diagram, so using a table is the better option.

	A	B	C	D	E	F
A	–	7	10	15	12	5
B	7	–	3	8	5	6
C	10	3	–	5	2	9
D	15	8	5	–	4	11
E	12	5	2	4	–	7
F	5	6	9	11	7	–

Example 2 is continued on the next page.

EXAMPLE 2 (CONT.)

b This table shows the tours produced by the nearest neighbour algorithm using each vertex in turn as the starting vertex.

Start vertex	Tour(s)	Total weight
A	AFBCEDA	35
B	BCEDFAB	32
C	CEDBFAC	35
D	DECBFAD	35
E	ECBFADE	35
F	FABCEDF	32

Check that you can see where these results come from.

The best tour found is 32 minutes.
With A as its starting point, it is ABCEDFA.
On the original network this corresponds to ABCEDEFA, because the route from D to F passes through E.

c On the original network there is a tour ABCDEFA = 31, so the solution found in **b** is not optimal.

Exercise 5.3

1 **a** Draw the complete network of shortest distances corresponding to the network shown.

b Show that the nearest neighbour algorithm with the four possible starting vertices leads to two distinct Hamiltonian tours, each with the same total weight.

c List the order in which the vertices would be visited on the original network, starting from A.

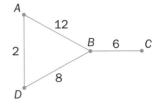

2 **a** Use the nearest neighbour algorithm with A as the starting vertex to obtain a Hamiltonian tour of the network shown. State the length of the tour.

b Show that by using B as the starting vertex you obtain a shorter tour.

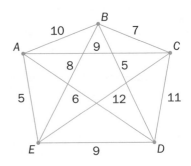

3 These are shortest routes (in km) between six Somerset towns: Cheddar (*C*), Frome (*F*), Glastonbury (*G*), Radstock (*R*), Shepton Mallet (*S*) and Wells (*W*). A lorry needs to travel from a warehouse in Radstock and make deliveries to shops in the other five towns.

	C	F	G	R	S	W
C	–	38	20	26	24	13
F	38	–	34	12	20	27
G	20	34	–	32	15	14
R	26	12	32	–	16	20
S	24	20	15	16	–	10
W	13	27	14	20	10	–

a Use the nearest neighbour algorithm with Radstock as the starting vertex to find a possible route, and state its length.

b Show that using Frome as the starting vertex leads to a shorter route. List the order in which the lorry visits the towns using this solution.

4 A guide is taking a party of tourists from a hotel (*H*) to four attractions in a city – the museum (*M*), the art gallery (*A*), the cathedral (*C*) and the Guildhall (*G*). She estimates the walking times (in minutes) between the various locations as shown in this network.

a Draw a complete network of shortest journey times.

b Use the nearest neighbour algorithm to plan a route for the party. State the total walking time and the order in which the route passes through the vertices of the original network.

5 The table shows the direct distances (in miles) between four locations. There are no other direct links between them.

a Fill in the blank cells in the table with the shortest indirect routes available.

	A	B	C	D
A	–	5		6
B	5	–		7
C			–	3
D	6	7	3	–

b Use the nearest neighbour algorithm with *A* as the starting vertex to find a Hamiltonian tour of your completed network. State the total length of the tour and list the order in which it would actually pass through the locations.

6 An orchestra and chorus is available as a whole unit (option *A*) or as four subgroups (*B*, *C*, *D* and *E*) offering smaller scale performances. The organiser of a six-day music festival wants to have a grand start and finish with option *A*, but to have the other four options appearing on the remaining four days. Because of travelling expenses etc. as performers come and go, the changeover costs (in £) between the options vary, as shown in this table.

	A	B	C	D	E
A	–	250	300	150	400
B	250	–	120	360	220
C	300	120	–	260	170
D	150	360	260	–	290
E	400	220	170	290	–

Advise the organiser as to the least expensive order in which to schedule the options.

D2

Because there is no efficient algorithm for finding the optimal solution to a TSP, you need a way of knowing what a 'good' solution is like. You calculate two values – an upper bound and a lower bound – such that

> lower bound ⩽ length of optimal tour ⩽ upper bound

The closer together the two bounds are, the more accurately you know the length of the optimal tour. You therefore try to find the smallest upper bound you can.

You also find the largest lower bound you can – see Section 5.5.

If you have found a tour, it is either optimal or longer. It follows that

> the length of any known tour is an upper bound.

So using the nearest neighbour algorithm to find a tour automatically provides an upper bound, but not necessarily a good one.
An alternative method makes use of a minimum spanning tree for the network. You could visit every vertex by travelling once in each direction along every edge of the minimum spanning tree, so

> (2 × length of minimum spanning tree) is an upper bound.

You can use this approach to find an upper bound without necessarily using the complete network of shortest distances, though the result may not be as good if the network does not satisfy the triangle inequality.

This is not usually a good upper bound, but you can improve it by using short cuts.

EXAMPLE 1

Use the minimum spanning tree (MST) and short cuts to find an upper bound for the optimal solution of the TSP for this network.

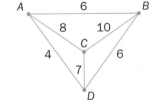

Use Kruskal's or Prim's algorithm to find two minimum spanning trees (MSTs), each of total length 17:

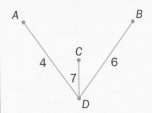 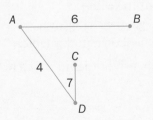

This gives an upper bound of 2 × 17 = 34

EXAMPLE 1 (CONT.)

For the first MST this upper bound is equivalent to visiting the vertices in the order *ADBDCDA*.

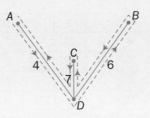

Improve this by replacing *BDC* (6 + 7 = 13) by *BC* (10):

This is a reduction of 13 − 10 = 3

So an improved upper bound is
34 − 3 = 31

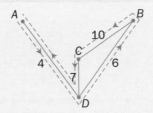

Improve this by replacing *CDA* (7 + 4 = 11) by *CA* (8):

This is a reduction of 11 − 8 = 3

So an improved upper bound is
31 − 3 = 28

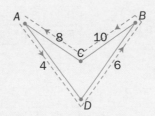

This is a tour, *ADBCA*, so there are no more short cuts. The best upper bound using this MST is 28.

Using the second MST, the initial upper bound of 34 is equivalent to visiting the vertices in the order *BADCDAB*.

Improve this by returning directly from *C* to *B*, replacing *CDAB* (7 + 4 + 6 = 17) by *CB* (10):

This is a reduction of 17 − 10 = 7
The improved upper bound is 34 − 7 = 27
You now have a tour *BADCB*,
so there are no more short cuts.
The best upper bound using this *MST* is 27.

You have a choice of 28 and 27 as the upper bound. You always want as small an upper bound as possible, so you have:

upper bound = 27

It is usually convenient to simplify the layout of the MST.

D2

BADCB = 27 is an optimal tour. However, the method will not achieve this in every case.

EXAMPLE 2

Find a minimum spanning tree for the network represented by this table. Use short cuts to find an upper bound for the optimal solution of the TSP for this network.

	A	B	C	D	E
A	–	6	7	2	12
B	6	–	4	8	9
C	7	4	–	5	7
D	2	8	5	–	10
E	12	9	7	10	–

Use Prim's algorithm to find the minimum spanning tree (MST):

The edges of the MST are BC, CD, DA and CE.

The total weight of the MST is 18. An initial upper bound is given by
$$2 \times 18 = 36$$

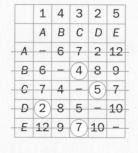

1	4	3	2	5
	A	B	C	D	E
A	–	6	7	2	12
B	6	–	4	8	9
C	7	4	–	5	7
D	2	8	5	–	10
E	12	9	7	10	–

This is equivalent to visiting the vertices in the order BCDADCECB.

Two possible short cuts are to go directly from A to E and from E to B.
The first replaces ADCE $(2 + 5 + 7 = 14)$ by AE (12).
This is a reduction of 2.
The second replaces ECB $(7 + 4 = 11)$ by EB (9).
This is a reduction of 2.

Using both of these improves the upper bound to
$$36 - 2 - 2 = 32$$

This is now a tour BCDAEB, so there are no more short cuts.

The final upper bound is 32.

This is not the length of the optimal tour, which is 29. Try finding a tour of this length. Can you also find a way of using short cuts on the original MST to give an upper bound of 29?

Exercise 5.4

1 The network shown represents the travelling times, in minutes, between four towns.

   a Find the minimum spanning tree for the network and hence state an upper bound for the optimal solution to the travelling salesman problem for the network

   b By introducing two short cuts, obtain an improved upper bound.

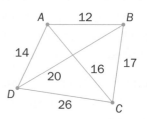

D2

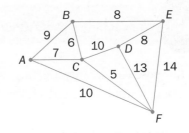

2 The network shows the distances, in km, between six towns. A van based at A needs to deliver to the other five towns and return to A.

a By finding the minimum spanning tree for the network, obtain an upper bound for the total distance that the van will need to travel.

b Use short cuts to show that the optimal tour is at most 50 km.

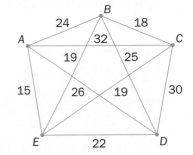

3 a Find the two minimum spanning trees for the network shown.

b Using the total weight of the minimum spanning tree, state an upper bound for the optimal solution to the TSP.

c Using short cuts, obtain the best upper bound you can using your minimum spanning trees.

4 a Use Prim's algorithm to find the minimum spanning tree for this network.

b State an upper bound for the solution to the TSP for this network.

c Using short cuts, obtain an upper bound less than 45.

	A	B	C	D	E	F
A	–	9	4	7	11	8
B	9	–	12	6	8	14
C	4	12	–	6	10	7
D	7	6	6	–	5	8
E	11	8	10	5	–	13
F	8	14	7	8	13	–

5 The network shows the direct distances, in miles, between eight villages. A van leaves the main post office at A and makes a round trip to collect mail from sub-post offices in B, C, D, E and F (G and H have no post office).

a Draw up a table of shortest distances between villages A–F.

b Show that there are two minimum spanning tree for your table from part a.

c State an upper bound for the distance the van needs to travel.

d Use short cuts to show that the van can do the job in less than 50 miles.

e State a possible route the van could take. How many times will it pass through G and H?

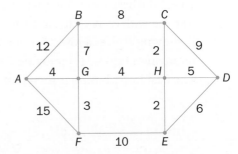

Suppose you are investigating this network. As well as an upper bound, you need a lower bound such that

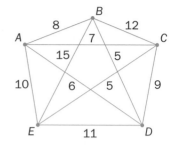

> lower bound ≤ length of optimal tour ≤ upper bound

The closer together the two bounds are the better.
You therefore try to find the largest lower bound you can.

The optimal tour enters and leaves *A* along two of *AB*, *AC*, *AD* and *AE*.

Separate these edges from the rest of the network:

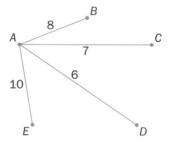

The two edges involved in the tour must total at least $6 + 7 = 13$

The remaining three edges of the tour join *B*, *C*, *D* and *E*. They form a spanning tree for the subgraph shown.

The minimum spanning tree for this subgraph is $5 + 5 + 9 = 19$, so the three edges must have a total weight of at least 19.

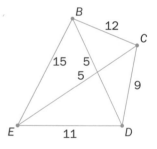

The complete tour must therefore have a total weight of at least $13 + 19 = 32$, so this is a lower bound for the tour.

To get the best lower bound available, repeat this process, starting with each vertex in turn.
Removing *B* gives a lower bound of 31, removing *C* gives 33, removing *D* gives 31, removing *E* gives 33.
The best lower bound is therefore 33.

Check that you can see where these values come from.

| The largest lower bound you find is the best one.

To find a lower bound:

Step 1 Choose a vertex *V*.

Step 2 Identify the two lowest weights, *p* and *q*, of the edges connected to *V*.

Step 3 Remove *V* and its connecting edges from the network. Find the total weight, *m*, of the minimum spanning tree of the remaining subgraph.

Step 4 Calculate lower bound = $p + q + m$

Step 5 If possible, choose another vertex *V* and go to Step 2.

Step 6 Choose the largest of the results obtained as the best lower bound.

Sometimes the minimum spanning tree and the two shortest edges form a Hamiltonian cycle. In this case what you have is a possible tour (and therefore an upper bound) which is also a lower bound. This means that this tour is the optimal tour.

D2

EXAMPLE 1

a By removing vertex *A*, obtain a lower bound for the solution to the travelling salesman problem for this network.

b By removing *B*, obtain a second lower bound. State which is the better value.

c Make a further deduction from your results for part **b**.

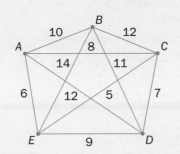

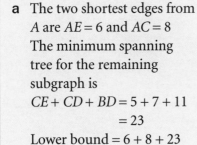

a The two shortest edges from *A* are $AE = 6$ and $AC = 8$
The minimum spanning tree for the remaining subgraph is
$$CE + CD + BD = 5 + 7 + 11$$
$$= 23$$
Lower bound $= 6 + 8 + 23$
$$= 37$$

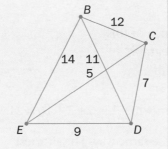

b The two shortest edges from *B* are $BA = 10$ and $BD = 11$
The minimum spanning tree for the remaining subgraph is
$$CE + AE + CD = 5 + 6 + 7$$
$$= 18$$
Lower bound $= 10 + 11 + 18$
$$= 39$$

The better value is the larger, 39.

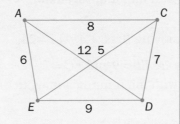

c The edges used in part **b** are *AB*, *BD*, *DC*, *CE* and *EA*, which form a tour, as shown. Because the length of this tour equals a known lower bound, it must be an optimal tour.

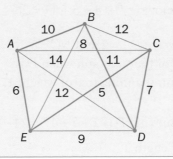

D2

EXAMPLE 2

In Section 5.4, Example 2, you found an upper bound of 32 for the TSP for this network.

Obtain the best the lower bound and state an inequality regarding the length of the optimal tour.

	A	B	C	D	E
A	–	6	7	2	12
B	6	–	4	8	9
C	7	4	–	5	7
D	2	8	5	–	10
E	12	9	7	10	–

Start by removing vertex A. The shortest edges are $AB = 6$ and $AD = 2$

In Section 5.4 you found the minimum spanning tree for this network, as shown.

Vertex A is at the end of a branch of this tree, so removing A gives a subgraph with minimum spanning tree of weight 16, as shown.

The lower bound found by removing A is $6 + 2 + 16 = 24$

Vertex B can be dealt with in the same way. The shortest edges are $BA = 6$ and $BC = 4$

Removing B gives a subgraph with minimum spanning tree of weight 14, as shown.

The lower bound found by removing B is $6 + 4 + 14 = 24$

For vertex E the shortest edges are $EB = 9$ and $EC = 7$

Removing E gives a subgraph with minimum spanning tree of weight 11, as shown.

The lower bound found by removing E is $9 + 7 + 11 = 27$

For vertex C the shortest edges are $CB = 4$ and $CD = 5$

C is not at the end of a branch of the complete MST, so find the MST of the subgraph when C is removed. This gives a minimum spanning tree of weight 17, as shown.

The lower bound found by removing C is $4 + 5 + 17 = 26$

For vertex D the shortest edges are $DA = 2$ and $DC = 5$

Again you must find the MST of the subgraph when D is removed. This gives a minimum spanning tree of weight 17, as shown.

The lower bound found by removing D is $2 + 5 + 17 = 24$

Suppose the optimal tour has length T.

The possible lower bounds for T are 24, 24, 27, 26 and 24.

The best lower bound is 27.

You know (from the previous example) that 32 is an upper bound for T. It follows that $27 \leqslant T \leqslant 32$

It can be useful to find the minimum spanning tree for the whole network if you intend to use every vertex in your search for a lower bound. Removing a vertex V from the network corresponds to removing V from the MST provided that V is at the end of a branch of the tree.

Exercise 5.5

1 **a** By deleting vertex A from this network, obtain a lower
       bound for the optimal tour. Explain, by reference to the
       network, why a tour of this length is not possible.

   b By deleting vertex B, obtain another lower bound for the
       optimal tour and explain why this is in fact the length of
       the optimal tour.

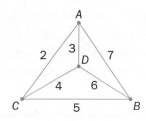

2 By deleting each vertex in turn from the network shown,
   obtain the best lower bound available for the solution
   to the travelling salesman problem.

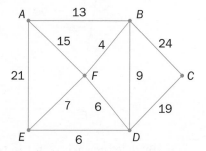

3 The diagram shows the locations of five nesting boxes
   in a bird reserve. The values shown are the lengths,
   in metres, of the various sections of path. A volunteer
   wants to visit every nest box, starting and finishing at A.

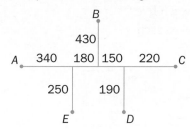

   a Draw a complete network of shortest distances.

   b The volunteer visits the nest boxes in alphabetical order.
       By deleting vertex A, find a lower bound for the TSP for
       this network and hence show that she could save no more
       than 120 m by finding the optimal route.

4 The table shows the distances, in km, between five locations.
   By deleting each vertex in turn, find the best lower bound
   for the solution to the travelling salesman problem for
   this network.

	A	B	C	D	E
A	–	14	22	18	12
B	14	–	15	32	25
C	22	15	–	20	13
D	18	32	20	–	28
E	12	25	13	28	–

D2

5 The diagram shows the cost, in $, of flying between five cities in the United States. A tourist arrives in and leaves the country at *A*, and wishes to travel to each of the cities in turn. Using the nearest neighbour algorithm the cheapest tour he can find is $420.

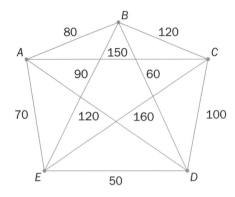

a By deleting each vertex in turn obtain the best available lower bound for the cost of his tour.

b State inequalities satisfied by the cost of the optimal route.

6 The table shows the direct distances, in km, between five Midland towns. A sales representative is to travel to each town in turn, starting and finishing in Coventry.

	Coventry	Nottingham	Leicester	Derby	Stoke on Trent
Coventry	–	77	37	12	93
Nottingham	77	–	40	24	80
Leicester	37	40	–	45	82
Derby	12	24	45	–	56
Stoke on Trent	93	80	82	56	–

By deleting each town in turn from the network, find the best lower bound for the distance she must travel.

7 **a** Draw up a table showing the shortest route between the vertices of the given network.

b Show that the best tour found by using the nearest neighbour algorithm exceeds the optimal tour by no more than 2.

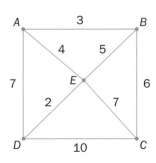

8 In an orienteering competition, contestants must visit each of five checkpoints once only and return to the start. They gain points depending on the difficulty of the route they follow – the harder the route the more points (but the greater the chance that they will not complete the course in the required time). The table shows the points to be gained on the various routes.

	Start	A	B	C	D	E
Start	–	10	8	14	9	12
A	10	–	5	6	18	15
B	8	5	–	10	12	7
C	14	6	10	–	15	6
D	9	18	12	15	–	10
E	12	15	7	6	10	–

a George wishes to travel the easiest route (least total points). Find the best lower and upper bounds for his optimal total, and state the best route that you can find for George to follow.

b Shahidar plans to gain as many points as possible (the hardest route). Find the best lower and upper bounds for her optimal total, and state the best route that you can find for Shahidar to follow.

9 A knight's move in chess consists of moving two squares parallel to one side of the board and then one square at right angles to this.

Belinda and Manuel play a game in which one of them places counters on six squares of the board and the other must make a 'knight's tour' starting and ending on one of the marked squares and capturing each of the other counters en route, making as few moves as possible.

The diagram shows the positions of the counters that Belinda has set up.

a Draw up a table to show the numbers of moves required to travel between each of these positions.

b Find upper and lower bounds for the number of moves that Manuel will have to make.

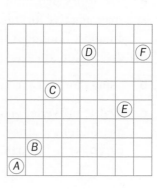

D2

1 **a** Draw a complete network of shortest distances corresponding to this network.

b Use the nearest neighbour algorithm on your complete network to find a possible solution to the travelling salesman problem. State the total weight of the tour you have found.

c List the order of the vertices in the corresponding route on the original network.

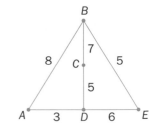

2 **a** Use the nearest neighbour algorithm to find a possible solution to the travelling salesman problem for this network.

b By deleting vertex A obtain a lower bound for the length of the optimal tour.

c By deleting vertex B obtain a second lower bound. State which of the two lower bounds is the better.

d Using your preferred lower bound and your answer to part **a**, write down inequalities for the length of the optimal tour.

3 The table shows the distances, in km, between seven locations. A distributor based at A needs to deliver bundles of newspapers to shops in each of the other locations.

a Use Prim's algorithm to find the minimum spanning tree for this network. Hence state an upper bound for the distance the distributor will need to travel.

	A	B	C	D	E	F	G
A	–	5	6	7	4	9	8
B	5	–	4	5	4	8	6
C	6	4	–	9	7	8	9
D	7	5	9	–	9	7	9
E	4	4	7	9	–	6	8
F	9	8	8	7	6	–	9
G	8	6	9	9	8	9	–

b Using short cuts, reduce your upper bound to below 50 km.

c By deleting vertex A obtain a lower bound for the distributor's journey. Using this and your result from part **b**, write inequalities satisfied by his optimal route.

D2

4 The table shows the travel times, in minutes, between six towns. A company intends to run a bus service on a circular route visiting all the towns.

	P	Q	R	S	T	U
P	–	10	20	16	18	11
Q	10	–	12	8	10	9
R	20	12	–	6	4	11
S	16	8	6	–	5	8
T	18	10	4	5	–	9
U	11	9	11	8	9	–

a Find the minimum spanning tree for the network. Using this and short cuts, show that the complete route can be completed in less than 50 minutes.

b By deleting P find a lower bound for the optimal route.

5 The network shows the distances, in miles, between five locations.

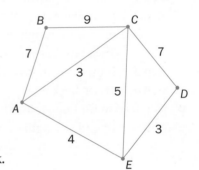

a Draw a complete network showing the shortest distances between the five locations.

b Use the nearest neighbour algorithm starting from A to obtain a possible solution to the travelling salesman problem. State the length of the route and the order in which the vertices would be visited on the original network.

c The nearest neighbour algorithm applied using the other starting vertices gives the following results: B, 32 miles, C, 33 miles, D, 33 miles, E, 31 miles. State the best upper bound for the optimal tour provided by this information.

d Show by reference to the original network that the nearest neighbour algorithm does not give an optimal tour.

e By deleting first A and then B from your complete network, obtain two lower bounds for the optimal tour.

f Write down inequalities giving the best range of values you have for the optimal tour.

D2

Summary

Refer to

- A Hamiltonian cycle (tour) is a closed path which visits every vertex of the graph. A Hamiltonian graph has at least one Hamiltonian cycle.

5.1

- The classical travelling salesman problem (TSP) is to visit every vertex of a network once only, and return to the start in the minimum distance.

5.2

- The nearest neighbour algorithm usually finds a reasonable, not necessarily optimal, solution:

  Step 1   Choose a starting vertex V.

  Step 2   From your current position choose the edge with minimum weight leading to an unvisited vertex. Travel to that vertex.

  Step 3   If there are unvisited vertices go to Step 2.

  Step 4   Travel back to V.

5.3

- You need to find upper and lower bounds for the optimal solution:
  - lower bound $\leqslant$ length of optimal tour $\leqslant$ upper bound

5.4, 5.5

D2

Links

The travelling salesman problem arises in many situations other than the obvious application.

In modern industrial processes, robots are often used to complete tasks.

Suppose that, in the manufacture of a circuit board, a robotic arm is required to solder a number of connections. The time the overall process takes depends on the total of the times taken for the robotic arm to move between each of the necessary solder points. By minimising the time taken for this process, the overall production time and the related costs can be kept to a minimum.

6

Network flows

This chapter will enable you to
- model the flow of a commodity through a network using a digraph
- identify a feasible flow
- improve on an initial flow by identifying a flow-augmenting path
- demonstrate that a flow is maximal using the maximum flow–minimum cut theorem.

Introduction

Many problems are concerned with the flow of a commodity through a network. The situation may involve the flow of oil through a network of pipes, the flow of traffic through a one-way system or the flow of data in a communications network.

In all these situations, there is a limit to the amount of the commodity that can flow along each part of the network. This chapter considers the problem of finding the maximum possible flow through the whole network.

D2

Suppose you have a directed network in which the edges are routes along which a commodity of some sort can flow. The weight of an edge is the maximum possible flow along that edge. This is the capacity of the edge. The network is sometimes called a capacitated network.

E.g.

The flow enters the network at S.

The flow leaves the network at T.

It is common to call the source S and the sink T. You will meet networks with more than one source or sink, but in this section there will be one of each.

At S all the edges are directed away from the vertex. A vertex like this is called a source.

At T all the edges are directed towards the vertex. A vertex like this is called a sink.

When the commodity flows through the network, there is a non-negative number – the flow in the edge – assigned to each edge. Together these form the flow in the network. The flow must satisfy certain conditions.

Firstly, the capacity of an edge is the maximum possible flow in that edge, so you have:

> The flow in an edge cannot be greater than its capacity.

This is sometimes called the feasibility condition.

Secondly, the commodity cannot be stored at a vertex, so you have:

> At every vertex apart from S and T
>   total inflow = total outflow

This is sometimes called the conservation condition.

It follows from this second condition that:

> total outflow from S = total inflow to T

This total outflow/inflow is called the value of the flow.

A feasible flow for the given network is:

At *A* inflow = outflow = 7
At *B* inflow = outflow = 9
Outflow at *S* = inflow at *T* = 12
The value of the flow = 12

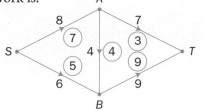

It is usual to circle the actual flows in the edges. So, in *SA*, for example, the flow is 7 and the capacity (maximum possible flow) is 8.

Notice that in *AB* and *BT* flow = capacity. These edges are said to be saturated. The other edges are unsaturated.

You usually need to find the maximum flow through the network. The network shown has a maximum flow of 14 (the most that can flow from *S*). Here is one way this flow could take place.

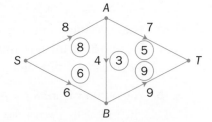

D2

EXAMPLE 1

The diagram shows a capacitated network. The circled values represent a feasible flow through the network.

a Find the values of *x* and *y*.

b State the value of the flow.

c State which edges are saturated.

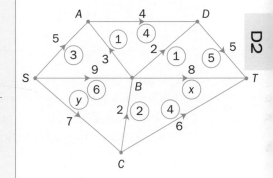

a At *B* the incoming edges are *SB* and *CB*, the outgoing edges *BA*, *BD* and *BT*.
By the conservation condition
   Total inflow = total outflow
   $6 + 2 = 1 + 1 + x$
   $x = 6$

At *C* the incoming edge is *SC*, the outgoing edges are *CB* and *CT*.
By the conservation condition
   $y = 2 + 4$
   $y = 6$

b The value of the flow is given by the total outflow from *S* or the total inflow to *T* (which should be equal).
   Outflow from $S = 3 + 6 + y = 15$
   Inflow to $T = 5 + x + 4 = 15$
so the value of the flow = 15

c An edge is saturated if its flow is the same as its capacity.

EXAMPLE 2

For the network shown

a find the maximum possible flow along the route *SABCT*

b state the value of the flow shown

c by considering the flow at three vertices, find the values of *x*, *y* and *z*

d by considering a fourth vertex, show that the flow shown is feasible.

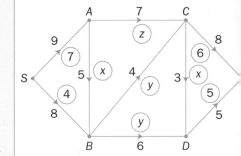

-----

a The capacities of the edges on the route *SABCT* are 9, 5, 4 and 8. As the flow must not exceed any of these, the maximum flow along this route is 4.

b The outflow from $S = 7 + 4 = 11$
The inflow to $T = 6 + 5 = 11$
The value of the flow = 11

c Apply the conservation condition at *A*, *B* and *C*:
    at *A* $x + z = 7$
    at *B* $x + 4 = 2y$
    at *C* $y + z = x + 6$
Solve these simultaneous equations: $x = 2, y = 3, z = 5$

d Consider the flow through *D*:
Total inflow $= x + y = 5$
Total outflow $= 5$
So *D* satisfies the conservation condition.

D2

Exercise 6.1

1 Showing all your working, confirm that the diagram shows a feasible flow through this network.

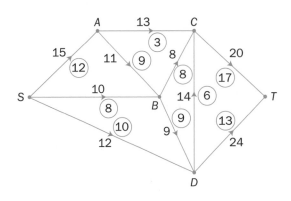

2 For the capacitated network shown

  a state the maximum possible flow along the route *SADCT*

  b state the value of the flow shown

  c find the values of *x*, *y* and *z*

  d explain why the vertex *B* is redundant.

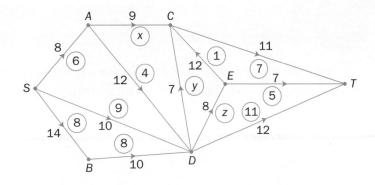

3 **a** For the capacitated network shown

    i find the values of *x* and *y*

    ii find the maximum possible flow along the route *SEBCT*.

  b Assuming that the flow in part **a ii** is taking place, what is then the maximum possible flow along the route *SDEBT*?

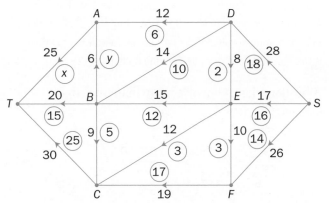

4 For the capacitated network shown

  a explain why *w* = 20

  b find the values of *x*, *y* and *z*

  c state which, if any, of the edges are saturated.

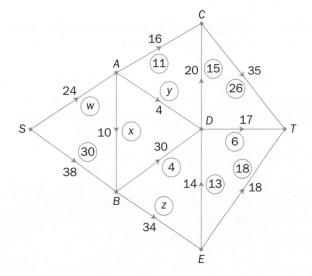

Traffic flow in a bottleneck of a road system is restricted and affects the overall flow through the system.

E.g.

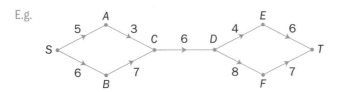

A flow of 9 is possible from S to C (3 along SAC, 6 along SBC).
A flow of 11 is possible from D to T (4 along DET, 7 along DFT).
But the maximum value of the flow in the network is only 6, which is the most that can pass through the bottleneck CD.

A bottleneck may involve more than one edge.

E.g.

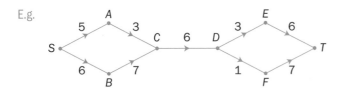

Here the bottleneck is formed by the edges DE and DF.
The maximum value of the flow is 4.

A bottleneck separates the network shown into two parts, one containing the source, S, and the other containing the sink, T.

The value of the network flow cannot be more than the flow through a bottleneck. The maximum value of the network flow is therefore given by the flow through the worst bottleneck.

This idea is formalised by the concept of a cut.

A cut is a set of edges whose removal disconnects the network into two parts X and Y, with X containing the source, S, and Y containing the sink, T.

Removing these edges would completely cut off the flow from S to T.

The capacity of a cut is the sum of the capacities of those edges of the cut which are directed from X to Y.

For a maximum network flow, edges directed from Y to X would need to have zero flow.

D2

You can describe a cut either by listing the set of edges in the cut, or by listing the vertices in the source set, X, and in the sink set, Y.

E.g.

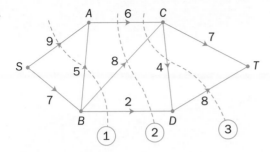

The diagram shows three possible cuts for this network.

1 The set of edges = {SA, BA, BC, BD}

Source set $X = \{S, B\}$, sink set $Y = \{A, C, D, T\}$

The capacity of the cut = $9 + 5 + 8 + 2 = 24$

The set of edges is sometimes called the **cut set**.

2 The set of edges = {AC, BC, BD}

Source set $X = \{S, A, B\}$, sink set $Y = \{C, D, T\}$

The capacity of the cut = $6 + 8 + 2 = 16$

3 The set of edges = {AC, BC, CD, DT}

Source set $X = \{S, A, B, D\}$, sink set $Y = \{C, T\}$

The capacity of the cut = $6 + 8 + 8 = 22$

The edge CD does not contribute to the capacity of cut 3 because it is directed from Y to X.

Any flow must cross from set X to set Y. It follows that

the value of any flow $\leqslant$ the capacity of any cut

The maximum flow corresponds to the worst bottleneck.

The maximum flow–minimum cut theorem
The value of the maximal flow = the capacity of a minimum cut

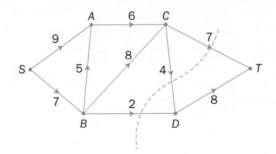

This diagram shows the minimum cut for this network.

The capacity of this cut = $2 + 4 + 7 = 13$

The maximum value of the flow is therefore 13.

D2

EXAMPLE 1

List all possible cuts for the network shown and find their capacities.

Hence state the maximal flow for the network.

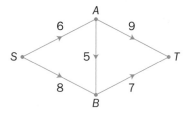

There are four possible cuts for this network, as shown.

1 {*SA, SB*} Capacity = 6 + 8 = 14

2 {*SA, AB, BT*} Capacity = 6 + 7 = 13

3 {*SB, AB, AT*} Capacity = 8 + 5 + 9 = 22

4 {*AT, BT*} Capacity = 9 + 7 = 16

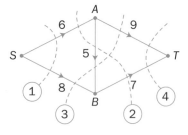

A useful consequence of the max flow–min cut theorem is:

> If you have a flow and a cut such that
>
> (value of flow) = (capacity of cut)
>
> then the flow is a maximum and the cut is a minimum.

The number of cuts increases rapidly for more complex networks. Even a simple network such as the one in Example 2 has nine possible cuts.

EXAMPLE 2

Find by inspection a minimum cut for the network shown. Confirm that it is a minimum cut by finding a flow with that value.

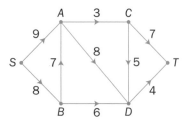

The edges in the required cut are likely to have small capacities. Look for the edges with the lowest weights.

In this case the cut {*AC, CD, DT*}, as shown, appears to be a minimum.

The capacity of this cut = 3 + 4 = 7

There is a flow of value 7, consisting of 3 along *SACT* and 4 along *SBDT*.

It follows that this flow is maximal, and the cut is a minimum.

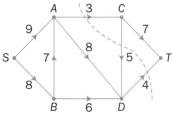

As before, *CD* goes the wrong way, so you could describe the cut as {*AC, DT*}.

Exercise 6.2

1 For each of the following diagrams state

   i the set of edges (the cut set)

   ii the source set X and the sink set Y

   iii the capacity of each of the cuts shown.

a

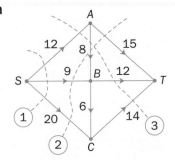

b

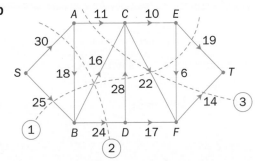

c

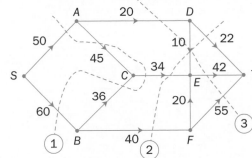

d

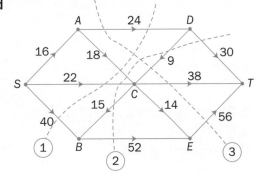

2 **a** For the network in question **1** part **a**

   i find a cut with a capacity of 35

   ii find a flow with a value of 35.

  b What can you deduce from your results in part **a**?

3 Find a minimum cut for the network in question **1** part **b**, and confirm that it is a minimum by finding a flow with that value.

4 For the network shown, find by inspection the maximum flow and a minimum cut.

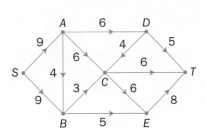

You need a systematic way of finding the maximal flow. Start with an initial flow of some sort and look for ways to augment (increase) it. This flow-augmentation process may happen several times until you have the maximal flow.

The initial flow could just be a zero flow, but you can usually find a better starting point.

The diagram shows a network with an initial flow of 3 along *SABT* and 2 along *SBT*.

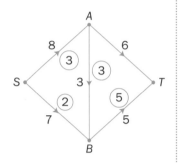

Subtract these flows from the capacities of the edges, to see how much spare capacity remains:

A flow of 5 is possible along *SAT*. *SAT* is called a flow-augmenting path.

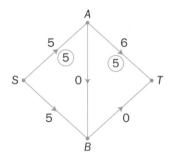

Subtract this flow to see what capacity remains:

There is no obvious flow-augmenting path, but you have not yet reached the maximal flow. If you increase the flow in *SB* and *AT* by 1, and *reduce* the flow in *AB* by 1, the overall flow increases by 1.

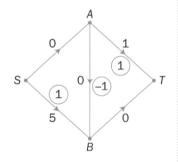

Effectively, 1 unit leaves *A* along *AT* instead of *AB*, and 1 unit enters *B* along *SB* instead of *AB*.

You have an overall flow of 11, as shown.

The edges *AT* and *BT* are saturated, so the flow is maximal.

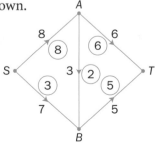

The labelling procedure

It is not easy to spot flow-augmenting paths, especially if they involve reducing a flow as in the network shown in this demonstration. It is also tedious to keep redrawing the diagram. The labelling procedure overcomes these problems.

Suppose an edge AB of capacity 12 has a flow of 8 along it. So far you have shown this as

$$A \xrightarrow{\quad 12 \quad} B$$
$$(8)$$

You could increase the flow by 4 – this is the potential flow. You could decrease the flow by 8 – this is the potential backflow. You show these as

$$A \xrightarrow[\longleftarrow 8]{\quad 4 \longrightarrow} B$$

The forward arrow represents the spare capacity.
The backward arrow represents the actual flow.
The total of the two gives the capacity of the edge.

The previous example now looks like this:

Set up the initial flow of 3 along the route $SABT$ and 2 along SBT:

The arrows pointing forward along the route SAT are $\geqslant 5$. SAT is a flow-augmenting path.

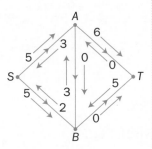

A flow-augmenting path is a route from S to T where the *arrows pointing forward* all have *non-zero values*. The lowest of these values gives the possible extra flow along that path.

Update the labels, reducing forward arrows and increasing backward arrows by 5:

The arrows pointing forward along the route $SBAT$ are all $\geqslant 1$, so this is a flow-augmenting path.

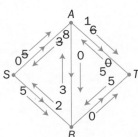

Update the labels by 1 along this route:

AT and BT are now saturated (potential flow = 0), so the flow is maximal. The potential backflow values (shown in blue) give the flows along the edges.

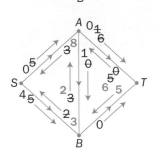

This is the stage at which previously you reduced the flow along AB by 1. The labelling procedure does this automatically.

This example is continued on the next page.

Using the other notation,
the maximal flow is as shown.

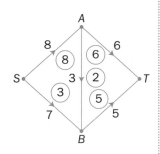

EXAMPLE 1

Taking a flow of 16 along *SACT* and 10 along
SBDT as the initial flow

a use flow-augmenting paths to find a
maximal flow for the network

b use the max flow–min cut theorem to
show that the flow is maximal.

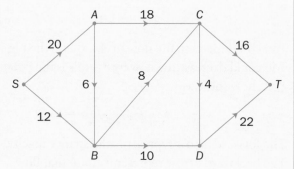

a Start by setting up the initial flow using the
standard labelling:

There is a flow-augmenting path *SACDT*,
with a potential flow of 2.

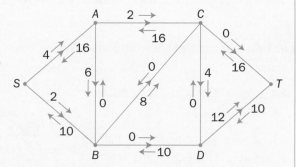

Update the labels to include this flow:

There is a flow-augmenting path *SBCDT*,
with a potential flow of 2.

All the forward arrows along *SBCDT*
have non-zero values.

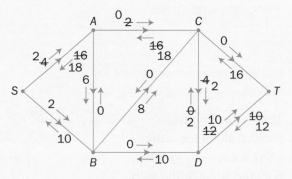

Update the labels to include this flow:

There is no flow-augmenting path available,
so the flow (shown in bold) is maximal.
The value of the flow is 30.

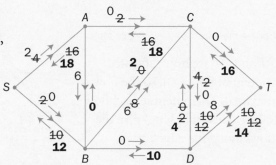

D2

EXAMPLE 1 (CONT.)

b There is a cut, {*BD*, *CD*, *CT*}, as shown, with capacity 30.

You have a flow of value 30 and a cut of capacity 30.
By the max flow–min cut theorem this flow is maximal.

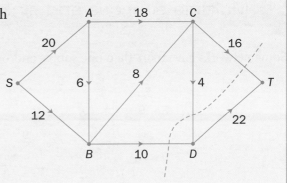

Exercise 6.3

1 Taking as the initial flow for the network shown a flow of 15 along *SAT*, 14 along *SBT* and 10 along *SCT*

   a use the labelling procedure to augment the flow until a maximal flow is obtained

   b confirm that your flow is maximal by using the maximum flow–minimum cut theorem.

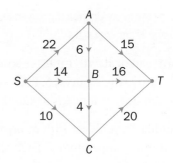

2 Taking as the initial flow for the network shown a flow of 20 along *SADT*, 20 along *SBCET* and 40 along *SBFT*

   a use the labelling procedure to augment the flow until a maximal flow is obtained

   b confirm that your flow is maximal by using the maximum flow–minimum cut theorem.

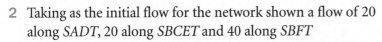
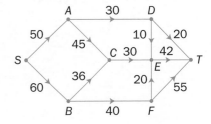

3 In the network shown, the maximum possible outflow from *S* and the maximum possible inflow to *T* are both 16. By starting with a flow of 5 along *SADT* and 5 along *SBET*, use flow augmentation to obtain a flow pattern with this maximum value.

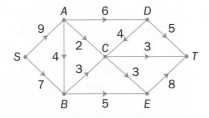

4 **a** For the network shown, use an initial flow and flow augmentation to find the maximal flow.

   b Use the max flow–min cut theorem to confirm that your flow is maximal.

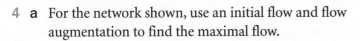
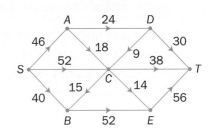

Some networks have more than one source and/or sink.

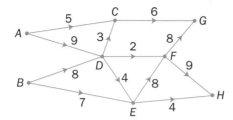

In this network there are no incoming edges at *A* or *B*.
A and *B* are both sources.

Similarly, there are no outgoing edges at *G* or *H*.
G and *H* are both sinks

You can deal with multiple sources by connecting them to a dummy **supersource**, *S*. Each source can receive from *S* as much flow as it needs to supply the network.

Similarly, you deal with multiple sinks by connecting them to a dummy **supersink**, *T*. Each sink can send to *T* all the flow it receives from the network.

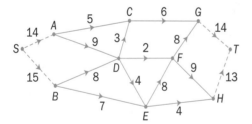

The possible outflow from *A* is 14.
The capacity of the dummy edge *SA* must be at least 14.
The possible outflow from *B* is 15.
The capacity of the dummy edge *SB* must be at least 15.

Similarly, the capacities of the dummy edges *GT* and *HT* must be at least 14 and 13 to cope with the possible inflows to *G* and *H*.

You can now find the maximal flow for the modified network by the usual methods. Once this is found, you can remove the dummy edges and vertices to leave the solution for the original network.

Exercise 6.4

1 The network shown has a sink *G* and two sources.

  a Identify the sources.

  b Introduce a supersource *S*.

  c Find the maximal flow from *S* to *G*.

  d Draw the original network to show the flow you have found, together with a minimum cut.

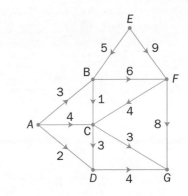

2 The network shown represents a system of one-way streets. Traffic enters the system at *A* and *B* and leaves at *I* and *H*. The weights are the maximum traffic flows, in hundreds of cars per hour, which can safely pass along the streets.

  a Introduce a supersource, *S*, and a supersink, *T*.

  b State the maximum possible flows along *ACFI*, *BDGJH* and *BEH*.

  c Using the flows from part **b** as your initial flow, use flow augmentation to obtain a network flow of value 19.

  d By finding a cut of capacity 19, show that the flow you found in part **c** is maximal.

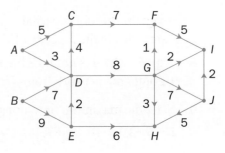

3 For the network shown

  a introduce a supersource and/or a supersink, as necessary

  b use an initial flow and flow augmentation to obtain a maximal flow

  c use the max cut–min flow theorem to confirm that the flow you have found is maximal.

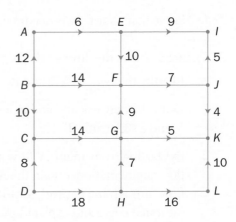

1

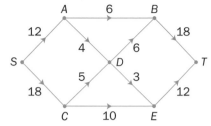

The diagram shows a capacitated network. The numbers on each arc indicate the capacity of that arc in appropriate units.

a Explain why it is not possible to achieve a flow of 30 through the network from S to T.

b State the maximum flow along

   i *SABT*   **ii** *SCET*.

c Show these flows on a copy of the diagram.

d Taking the answer to part **c** as the initial flow pattern, use the labelling procedure to find the maximum flow from S to T. List each flow-augmenting path you use, together with its flow.

e Indicate a maximum flow on a copy of the diagram.

f Prove that your flow is maximal.

[(c) Edexcel Limited 2001]

2 The table shows the flows in the edges of a network.

a Identify the two sources and two sinks.

b Draw the network and add in a supersource, S, and a supersink, T.

c By starting from an initial flow and using flow augmentation, obtain the maximal flow through the network. List the saturated edges and explain how they show that the flow is maximal.

From	To							
	A	B	C	D	E	F	G	H
A	–	12	8	–	–	–	–	–
B	–	–	–	10	–	14	–	–
C	–	11	–	–	–	–	12	–
D	–	–	–	–	–	–	–	–
E	–	–	9	–	–	–	15	–
F	–	–	8	15	–	–	–	7
G	–	–	–	–	–	15	–	18
H	–	–	–	–	–	–	–	–

3

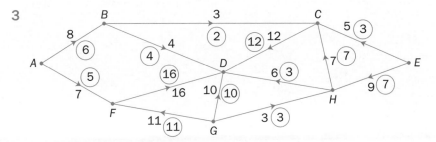

The network in this diagram models a drainage system.
The uncircled number on each arc indicates the capacity
of that arc, in litres per second. The circled numbers show
a feasible flow through the network.

a Write down the source vertices.

b State the value of the feasible flow shown.

c Taking the flow shown as your initial flow pattern, use the
labelling procedure to find a maximum flow through this
network. You should list each flow-augmenting route you
use, together with its flow.

d Show the maximal flow on a copy of the diagram and
state its value.

e Prove that your flow is maximal. [(c) Edexcel Limited 2002]

4 The diagram shows a network of
roads represented by arcs.
The capacity of the road represented
by that arc is shown on each arc.
The numbers in circles represent
a possible flow of 26 from B to L.

Three cuts C_1, C_2 and C_3 are shown
in the diagram.

a Find the capacity of each of the
three cuts.

b Verify that the flow of 26 is maximal.

The government aims to maximise the possible flow from B to L by
using one of two options.

    Option 1:   Build a new road from E to J with capacity 5.
    Option 2:   Build a new road from F to H with capacity 3.

c By considering *both* options, explain which one meets the
government's aim.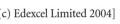

D2

5

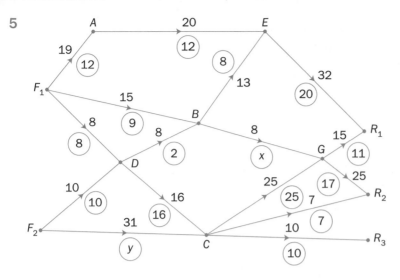

The diagram shows a capacitated, directed network of pipes flowing from two oil fields F_1 and F_2 to three refineries R_1, R_2 and R_3. The number on each arc represents the capacity of the pipe and the numbers in the circles represent a possible flow of 65.

a Find the value of x and the value of y.

b On a copy of the diagram, add a supersource and a supersink, and arcs showing their minimum capacities.

c Taking the given flow of 65 as the initial flow pattern, use the labelling procedure to find the maximum flow. State clearly your flow augmenting routes.

d Show the maximum flow on a copy of the diagram and write down its value.

e Verify that this is the maximum flow by finding a cut equal to the flow.

[(c) Edexcel Limited 2003]

6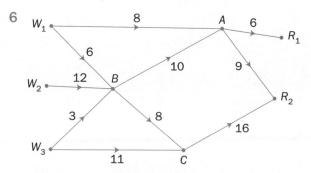

A company has three warehouses W_1, W_2, and W_3. It needs to transport the goods stored there to two retail outlets R_1 and R_2. The capacities of the possible routes, in van loads per day, are as shown.

Warehouses W_1, W_2 and W_3 have 14, 12 and 14 van loads respectively available per day and retail outlets R_1 and R_2 can accept 6 and 25 van loads respectively per day.

a   On a copy of the diagram, add a supersource W, a supersink R and the appropriate directed arcs to obtain a single-source, single-sink capacitated network. State the minimum capacity of each arc you have added.

b   State the maximum flow along

   i $W\,W_1\,A\,R_1\,R$
   ii $W\,W_3\,C\,R_2\,R$

c   Taking your answers to part **b** as the initial flow pattern, use the labelling procedure to obtain a maximum flow through the network from W to R. List each flow-augmenting route you use, together with its flow.

d   From your final flow pattern, determine the number of van loads passing through B each day.

[(c) Edexcel Limited 2002]

D2

Exit ⇒

Summary

Refer to

- A flow in a network is feasible if it satisfies two conditions:
  - The flow in an edge cannot exceed the capacity of that edge.
    If flow = capacity, the edge is saturated.
  - At every vertex (apart from S and T) total inflow = total outflow
    It follows that total outflow from S = total inflow to T
    This is the value of the flow.   6.1
- A cut is a set of edges whose removal disconnects the network
  into two parts X and Y, with X (the source set) containing
  S and Y (the sink set) containing T.
  - The maximum flow–minimum cut theorem states:
      the value of the maximal flow = the capacity of a minimum cut
    A given flow is maximal if you can find a cut whose
    capacity equals the value of the flow.   6.2
- To find a maximal flow you start with an initial flow and look
  for a flow-augmenting path, that is a route from S to T where
  all edges have spare capacity. You augment the initial flow by
  adding in this new flow. The process is made easier by using the
  labelling procedure.   6.3
- If a network has multiple sources (sinks), you modify it by connecting
  them all to a dummy supersource (supersink).   6.4

Links

Network flow problems arise in relation to
environmental management.

In an area at risk of flooding, it is important
to be able to calculate how rapidly the
existing drainage system can remove water.
If the drainage system is not effective enough
to prevent further flooding, then an improved
infrastructure needs to be designed.

7

Dynamic programming

This chapter will enable you to

- understand Bellman's principle of optimality
- use dynamic programming to solve multi-stage decision-making problems
- present your solution in graphical or tabular form
- understand the terms minimax and maximin when applied to routing problems.

Introduction

Dynamic programming enables you to analyse problems in which a series of decisions must be made.

For example, in drawing up a production plan for a period of three months, you might first decide how much to produce in the first month. This decision and its outcome will affect the next decision regarding the production for the second month, and so on.

Dynamic programming was introduced in the 1950s following publications by Richard Bellman, who was a mathematician with the Rand Corporation in the USA. It has since become a common tool in areas such as production and maintenance planning and stock control.

Suppose that you plan to cycle from Yeovil (*Y*) to Crawley (*Cr*), taking three days over the journey. You will stop at either Salisbury (*S*) or Ringwood (*R*) after the first day, and Fareham (*F*), Petersfield (*P*) or Chichester (*Ch*) after the second.

The table shows the distances in miles. You want to minimise the total distance.

		To					
		S	**R**	**F**	**P**	**Ch**	**Cr**
	Y	42	50	–	–	–	–
	S	–	–	51	48	56	–
From	**R**	–	–	54	47	50	–
	F	–	–	–	–	–	38
	P	–	–	–	–	–	40
	Ch	–	–	–	–	–	36

You can draw this as a directed network:

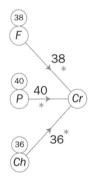

If you were solving this using Dijkstra's algorithm, you would start at Yeovil and work forwards. With dynamic programming you work backwards from the finish.

You could work forwards with dynamic programming, but backward working is the approach you will be expected to take in this unit.

At the first stage of the solution you are at Fareham, Petersfield or Chichester. These are the **stage 1 vertices**.

From Fareham, there is one possible route, length 38 miles. Label Fareham with 38, and asterisk the edge *F*–*Cr*. The asterisk shows which edge to take if the optimal route passes through Fareham.

Similarly, label Petersfield with 40 and Chichester with 36, asterisking both routes.

Now look at Salisbury and Ringwood, the **stage 2 vertices**.

The distance from *S* to *Cr* via *F* is given by
(edge *SF* + label of *F*) = 51 + 38 = 89 miles

Similarly, distance via *P* = 48 + 40 = 88 miles, and via *Ch* = 56 + 36 = 92 miles

The shortest is 88, so label *S* with 88 and asterisk the edge *S*–*P* to show which route to take.

Similarly, the best route from *R* is 86 miles via *Ch*, so label *R* with 86 and asterisk the edge *R*–*Ch*.

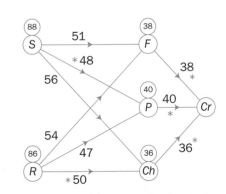

D2

Finally, look at Yeovil, the stage 3 vertex.

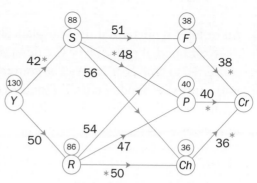

The distance from Y to Cr via S is given by
(edge YS + label of S) = 42 + 88 = 130 miles.

Similarly, distance via R = 50 + 86 = 136 miles

The shortest is 130, so label Y with 130 and asterisk
the edge Y–S to show which route to take.

This gives the shortest distance as 130 miles.
The route, Y–S–P–Cr, is shown by the asterisked edges.

In solving this problem you made an important assumption.
Having found that the best route from S is S–P–Cr = 88, you
assumed that the best route from Y via S would be Y–S–P–Cr.
You didn't need to consider Y–S–F–Cr or Y–S–Ch–Cr,
but just used the label you had given to S.
This is an example of Bellman's principle of optimality.

Bellman's principle of optimality
Any part of an optimal path is itself optimal.

In using dynamic programming you need to be familiar
with some notation and vocabulary.
Consider this network:

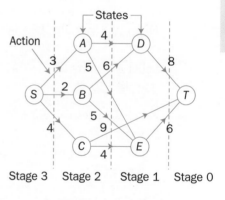

Each vertex is a state, a situation in which a decision must be
made. It is common to label the initial state S and the end state T.

Each state belongs to a stage, corresponding to how far it is
from the end of the process.

A state V belongs to stage n if routes from V to the end state
have a maximum of n edges.

Each vertex can be identified by its stage and state. You write
(stage; state), so for example vertex A is the first state in stage 2
and so is $(2; 1)$. Similarly C is $(2; 3)$.

At each vertex you need to decide between a number of edges.
These are called actions. The weight on the edge is a value
associated with the action. Depending on the nature of the
problem it may be a distance, a time, a profit/cost or some
other quantity.

Vertex C is a stage 2 vertex even
though there is a direct route to
T, because there is a route C-E-T
with two edges.

An investor is offered a three-year plan. She can buy a one-year bond of type *A*, *B* or *C*. After a year, *A* and *B* can be converted to type *D* or *E*. *C* can be converted to type *E* or put into a two-year bond of type *T*. *D* and *E* convert to *T* after a further year. The diagram shows this structure and the profit, in £000, which each bond will pay. What investment plan should she follow to maximise her total profit?

The stage 1 vertices are *D* and *E*.

These give profits of 8 and 6 respectively. They are shown labelled and asterisked in the diagram.

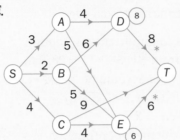

Work backwards from T.

The stage 2 vertices are *A*, *B* and *C*.

At *A*, she can choose *D* or *E*.
D: profit = (weight *AD* + label of *D*)
       = 4 + 8 = 12
E: profit = (weight *AE* + label *E*)
       = 5 + 6 = 11
Label *A* = max (12, 11) = 12
Asterisk *AD*.

Label *B* = max(6 + 8, 5 + 6) = 14
Asterisk *BD*.
Label *C* = max (9, 4 + 6) = 10
Asterisk *CE*.

The stage 3 vertex is *S*.
Label *S* = max (3 + 12, 2 + 14, 4 + 10) = 16
Asterisk *SB*.

Her maximum profit is £16 000. The asterisks show the route is *SBDT*. She should buy a type *B* bond and then convert to a type *D* bond.

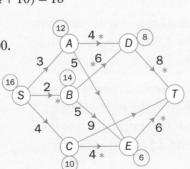

Exercise 7.1

1 Use dynamic programming to find the shortest route through this network from S to T. State the route and its length.

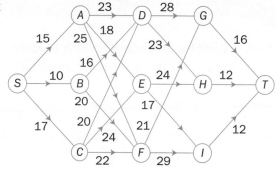

2 Repeat question **1** to find the longest route from S to T.

3 A production line is to be used to make three products P, Q and R, each for a one-week period. The efficient use of labour and materials means that the cost (in £000) for each product depends on the week in which it is made. The table shows the possible costs, and the network illustrates the possible orders in which the products might be made.

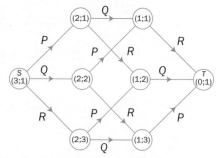

a Copy the network and label the edges with the costs.

b Use dynamic programming to find the minimum cost route through the network.

c State the order in which the products should be made and the total cost involved.

Cost in £000			
	Week		
		1	2	3
Product				
P		15	20	19
Q		30	28	24
R		21	22	24

4 Colin is going to travel by bus to visit his sister Colette. He will need to change buses at A, B or C and again at D, E or F. The table shows the ticket prices (in £) between the various locations.

a Draw a directed network to show the possible stages of his journey.

b Show, using dynamic programming, that Colin has a choice of two least-cost routes. State the possible routes and what his total cost will be.

		To						
		A	B	C	D	E	F	Colette's house
	Colin's house	7	5	8	–	–	–	–
	A	–	–	–	9	9	6	–
	B	–	–	–	4	8	5	–
From	C	–	–	–	6	6	8	–
	D	–	–	–	–	–	–	8
	E	–	–	–	–	–	–	6
	F	–	–	–	–	–	–	7

D2

Using a table

In more complex examples it is easier to show your calculations in a table. To illustrate this, consider the cycle trip from Yeovil to Crawley as in Section 7.1.

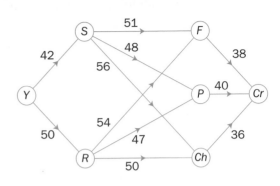

For stage 1 the table shows the states *F*, *P* and *Ch* and the possible actions. It reproduces the information you showed on the network.

Stage	State	Action	Destination	Value
	F	F–Cr	Cr	38*
1	P	P–Cr	Cr	40*
	Ch	Ch–Cr	Cr	36*

At stage 2 the table shows the states *S* and *R* and the possible actions.

Unlike before, show the consequence of every action, and asterisk the best one:

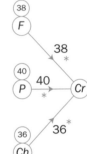

Stage	State	Action	Destination	Value
	F	F–Cr	Cr	38*
1	P	P–Cr	Cr	40*
	Ch	Ch–Cr	Cr	36*
		S–F	F	51 + 38 = 89
	S	S–P	P	48 + 40 = 88*
		S–Ch	Ch	56 + 36 = 92
2		R–F	F	54 + 38 = 92
	R	R–P	P	47 + 40 = 87
		R–Ch	Ch	50 + 36 = 86*

Finally, complete the table to show the choices at stage 3. You can then read back through the table, following the asterisks to find the optimal route (shown in blue).

Stage	State	Action	Destination	Value
	F	F–Cr	Cr	38*
1	P	P–Cr	Cr	40*
	Ch	Ch–Cr	Cr	36*
		S–F	F	51 + 38 = 89
	S	S–P	P	48 + 40 = 88*
		S–Ch	Ch	56 + 36 = 92
2		R–F	F	54 + 38 = 92
	R	R–P	P	47 + 40 = 87
		R–Ch	Ch	50 + 36 = 86*
3	Y	Y–S	S	42 + 88 = 130*
		Y–R	R	50 + 86 = 136

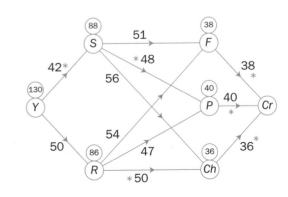

EXAMPLE 1

A small firm makes hand-built dining room suites. They can make at most 5 per month and can store up to 3 from one month to the next. There is a fixed cost of £500 if they work in a given month and each suite costs £800 to produce. Storage costs £200 per suite. The order book for the next four months is:

Month	Mar	Apr	May	Jun
Demand	1	3	2	4

They start with none in stock and want to end the four-month period with none in stock. Find the optimal production schedule.

Stage 1 is the final month, June. You must end the month with no stock. To satisfy a demand of 4 you could, for example, start with 2 in stock and make 2 more. This has a value (cost) of £500 + 2 × £800 = £2100 The table shows all the possibilities.

Stage	Demand	State (in stock)	Action (no. to make)	Destination (closing stock)	Value (×£100)
1 (Jun)	4	0	4	0	37*
		1	3	0	29*
		2	2	0	21*
		3	1	0	13*

As June can start with between 0 and 3 in stock, these are the possible closing stocks for May (stage 2). May can start with between 0 and 3 in stock, and demand is 2.

Extend the table to show the possible situations:

For example, if you started with 1 in stock and made 4, you would need to store 3. The value of this would be £500 + 4 × £800 + 3 × £200 = £4300.

This would mean June started with 3, with a value of £1300. The total value of this option is therefore £4300 + £1300 = £5600.

Example 1 is continued on the next page.

EXAMPLE 1 (CONT.)

Stage	Demand	State (in stock)	Action (no. to make)	Destination (closing stock)	Value (×£100)
1 (Jun)	4	0	4	0	37*
		1	3	0	29*
		2	2	0	21*
		3	1	0	13*
2 (May)	2	0	2	0	21 + 37 = 58*
			3	1	31 + 29 = 60
			4	2	41 + 21 = 62
			5	3	51 + 13 = 64
		1	1	0	13 + 37 = 50*
			2	1	23 + 29 = 52
			3	2	33 + 21 = 54
			4	3	43 + 13 = 56
		2	0	0	0 + 37 = 37*
			1	1	15 + 29 = 44
			2	2	25 + 21 = 46
			3	3	35 + 13 = 48
		3	0	1	2 + 29 = 31*
			1	2	17 + 21 = 38
			2	3	27 + 13 = 40

Mark with an asterisk the optimal value for each stage and state.

Continuing in this way, list the possible options at stage 3 and stage 4:

Stage	Demand	State (in stock)	Action (no. to make)	Destination (closing stock)	Value (×£100)
1 (Jun)	4	0	4	0	37*
		1	3	0	29*
		2	2	0	21*
		3	1	0	13*
2 (May)	2	0	2	0	21 + 37 = 58*
			3	1	31 + 29 = 60
			4	2	41 + 21 = 62
			5	3	51 + 13 = 64
		1	1	0	13 + 37 = 50*
			2	1	23 + 29 = 52
			3	2	33 + 21 = 54
			4	3	43 + 13 = 56
		2	0	0	0 + 37 = 37*
			1	1	15 + 29 = 44
			2	2	25 + 21 = 46
			3	3	35 + 13 = 48
		3	0	1	2 + 29 = 31*
			1	2	17 + 21 = 38
			2	3	27 + 13 = 40

Check that you fully understand how this table has been obtained.

EXAMPLE 1 (CONT.)

3 (Apr)	3	0	3	0	$29 + 58 = 87$
			4	1	$39 + 50 = 89$
			5	2	$49 + 37 = 86*$
		1	2	0	$21 + 58 = 79$
			3	1	$31 + 50 = 81$
			4	2	$41 + 37 = 78*$
			5	3	$51 + 31 = 82$
		2	1	0	$13 + 58 = 71$
			2	1	$23 + 50 = 73$
			3	2	$33 + 37 = 70*$
			4	3	$43 + 31 = 74$
		3	0	0	$0 + 58 = 58*$
			1	1	$15 + 50 = 65$
			2	2	$25 + 37 = 62$
			3	3	$35 + 31 = 66$
4 (Mar)	1	0	1	0	$13 + 86 = 99*$
			2	1	$23 + 78 = 101$
			3	2	$33 + 70 = 103$
			4	3	$43 + 58 = 101$

At stage 4 you asterisked 99, so the optimal cost is £9900.
Follow the asterisks up the table (shown in blue) to find the optimal sequence of actions.

The optimal production plan is to make 1 suite in March, 5 in April, none in May and 4 in June.

Although it is optimal, this plan might not be realistic unless the firm makes other items as well. You probably wouldn't want to lay off your workforce for the month of May.

D2

Exercise 7.2

1

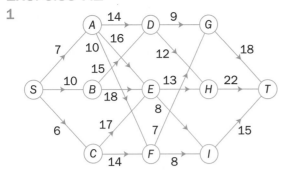

Use dynamic programming to find the shortest route through this directed network from *S* to *T*.
Show your working in the form of a table.

2

To From	A	B	C	D	E	F	G	T
S	27	25	28	24	–	–	–	–
A	–	–	–	–	16	23	14	–
B	–	–	–	–	20	21	26	–
C	–	–	–	–	22	15	28	–
D	–	–	–	–	24	18	20	–
E	–	–	–	–	–	–	–	29
F	–	–	–	–	–	–	–	26
G	–	–	–	–	–	–	–	19

Rough timber is taken from a port, *S*, to one of four mills, *A*, *B*, *C* or *D*, for cutting. It goes from there to one of three plants, *E*, *F* or *G*, for finishing, and finally to a distribution warehouse, *T*.

A lorry takes a load from the port to one of the mills, a load from there to one of the plants and then a load to the warehouse. The table shows the tonnage of the available loads. Showing your working in the form of a table, use dynamic programming to find the route which maximises the weight of freight transported.

3

Week \ City	1	2	3
A	210	240	280
B	190	280	270
C	200	270	230

The states in your table correspond to the cities already visited.

A tourist wants to spend three weeks in Spain, one week in each of Alicante (*A*), Barcelona (*B*) and Cordoba (*C*). His chosen hotels have prices which vary from week to week, as shown (in euros) in the table.

Showing your working in the form of a table, use dynamic programming to find the least expensive order in which he should visit the cities.

4 A firm makes sets of dog agility equipment. They can make a maximum of four sets per week and can store up to two sets from week to week. There are set-up costs of £50 each week (unless they make no sets that week). Each set costs £200 to make and it costs £30 per set per week for storage. They have nothing in stock at the start of a four-week period and want to end the period with no stock. They have orders for two, five, three and two sets in weeks 1, 2, 3 and 4 respectively.

Showing your working in the form of a table, use dynamic programming to find the least-cost production plan.

D2

Suppose you plan to walk from S to T, with an overnight stop at A or B and another at C, D or E, as shown in this network. The values are the distances in miles.

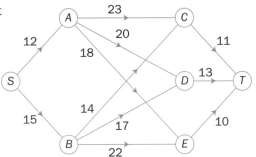

You want to avoid having to walk too far in one day, so you look for a route on the network in which the longest edge is as short as possible. This is called the minimax route.

> A minimax route is one in which the maximum edge weight is as small as possible – it minimises the maximum weight.

You can show the working on the network or in a table – both layouts are shown here. The values recorded are the maximum edge weights used.

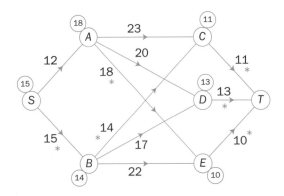

Check that you fully understand how both layouts have been obtained.

Stage	State	Action	Destination	Value
1	C	C–T	T	11*
	D	D–T	T	13*
	E	E–T	T	10*
2	A	A–C	C	max(23, 11) = 23
		A–D	D	max(20, 13) = 20
		A–E	E	max(18, 10) = 18*
	B	B–C	C	max(14, 11) = 14*
		B–D	D	max(17, 13) = 17
		B–E	E	max(22, 10) = 22
3	S	S–A	A	max(12, 18) = 18
		S–B	B	max(15, 14) = 15*

In both layouts you can see that the minimax route is $SBCT$. The maximum day's journey is 15 miles.

D2

In the same way, you sometimes need to find the route for which the minimum weight involved is as large as possible. This is called the maximin route.

> A maximin route is one in which the minimum edge weight is as large as possible – it maximises the minimum weight.

EXAMPLE 1

The diagram shows a number of roads. The weight shown on each edge indicates the heaviest lorry, in tonnes, allowed to pass along the road. Use dynamic programming to find the heaviest lorry able to travel from S to T, and find the route that it should take.

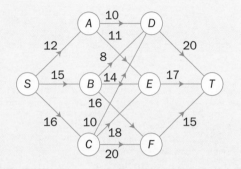

You can show all the working on the network or as a table. Both approaches are illustrated here:

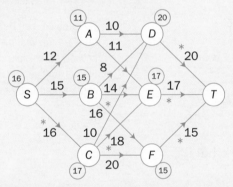

Check that you fully understand how both layouts have been obtained.

Stage	State	Action	Destination	Value
1	D	D–T	T	20*
	E	E–T	T	17*
	F	F–T	T	15*
2	A	A–D	D	min(10, 20) = 10
		A–E	E	min(11, 17) = 11*
	B	B–D	D	min(8, 20) = 8
		B–E	E	min(14, 17) = 14
		B–F	F	min(16, 15) = 15*
	C	C–D	D	min(10, 20) = 10
		C–E	E	min(18, 17) = 17*
		C–F	F	min(20, 15) = 15
3	S	S–A	A	min(12, 11) = 11
		S–B	B	min(15, 15) = 15
		S–C	C	min(16, 17) = 16*

The heaviest lorry is 16 tonnes. The optimal route is *SCET*.

EXAMPLE 2

A college plans three building projects, one in each of the next three years. Costs change depending on which buildings have already been completed (because of problems of access and installing utilities). The table shows the costs (in units of £10 000). They want the largest annual outlay to be as small as possible.

Project Already built	A	B	C
None	12	15	10
A	–	18	14
B	14	–	16
C	13	17	–
A and B	–	–	15
A and C	–	16	–
B and C	17	–	–

You need to minimise the maximum outlay, so you want the minimax route. You can model the problem as a network:

The problem can be solved in tabular form without drawing a network. In complex situations the network can be too complicated to be helpful.

You could show your working on the network, but it is shown here in tabular form.

Stage	State (Projects completed)	Action	Destination	Value
1	(1; 1) AB	Build C	(0; 1) T	15*
	(1; 2) AC	Build B	(0; 1) T	16*
	(1; 3) BC	Build A	(0; 1) T	17*
2	(2; 1) A	Build B	(1; 1)	max(18, 15) = 18
		Build C	(1; 2)	max(14, 16) = 16*
	(2; 2) B	Build A	(1; 1)	max(14, 15) = 15*
		Build C	(1; 3)	max(16, 17) = 17
	(2; 3) C	Build A	(1; 2)	max(13, 16) = 16*
		Build B	(1; 3)	max(17, 17) = 17
3	(3; 1) S None	Build A	(2; 1)	max(12, 16) = 16
		Build B	(2; 2)	max(15, 15) = 15*
		Build C	(2; 3)	max(10, 16) = 16

Check that you fully understand how these results have been obtained.

The optimal order is to build B, then A, then C.
The largest annual outlay will be £150 000.

Exercise 7.3

1 The network shows the distances, in km, between a number of towns. A cyclist wishes to travel from S to T, making two stops on the way. She wants to choose her route so that the longest leg of her journey is as short as possible.

a Copy the network and find the cyclist's optimal route. Show all your working on the diagram and state the length of the longest leg of her journey.

b Is the optimal route in part **a** also the shortest route?

2 Find the maximin route for the network in question **1**. Show your working in tabular form.

3 The diagram shows the weight restrictions, in tonnes, on the roads between S and T. Showing your working in tabular form, use dynamic programming to find the heaviest vehicle which can legally travel from S to T. State the route that it should take.

4 A company has three projects to complete. It can only do one at a time. The profit from each project depends on which projects have already been completed. The profits, in £000, are shown in the table.

In order to have as regular an income as possible, the company wishes to complete the projects so that the minimum profit gained is as great as possible.

a Draw a network to model this situation.

b Showing your working in tabular form, use dynamic programming to find the optimal order in which to complete the projects.

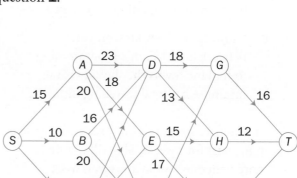

Completed \ Project	A	B	C
None	7	6	9
A	–	10	8
B	9	–	12
C	10	8	–
A and B	–	–	11
A and C	–	10	–
B and C	6	–	–

1 Belén lives in a rural area, so she has to drive to specialist shops. She wants to buy a settee and then go straight on to get a carpet to match. Settees are on sale in towns A, B and C, and carpets in towns D, E and F. The table shows the distances in miles between the various locations.

		To						
		A	B	C	D	E	F	Belén's house
From	Belén's house	10	6	9	–	–	–	–
	A	–	–	–	8	9	7	–
	B	–	–	–	5	8	12	–
	C	–	–	–	10	6	9	–
	D	–	–	–	–	–	–	13
	E	–	–	–	–	–	–	11
	F	–	–	–	–	–	–	14

a Draw a directed network to show the possible stages of her journey.

b Showing your working on your network, use dynamic programming to find the towns she should visit to minimise her total travelling distance.

2 An engineer needs to visit three overseas sites in the next four weeks. She will not make more than one visit in any given week. The cost of air travel fluctuates, so the potential costs (in £) are as shown in the table.

	Week 1	Week 2	Week 3	Week 4
Site A	180	180	200	175
Site B	90	140	130	130
Site C	170	190	180	160

a Draw a directed network, with the states corresponding to the sites, if any, that have been visited, to model this situation.

b Showing your working in tabular form, use dynamic programming to find the optimal schedule of visits.

3 **a** Use dynamic programming to find the minimax route for the network shown. Show your working on a copy of the diagram.

b Use dynamic programming to find the maximin route for the network shown. Show your working in tabular form.

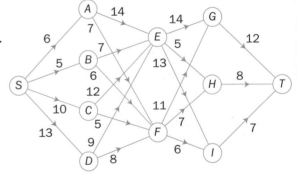

4 A company makes greenhouses.
The table shows its order book for the
next four weeks. It has no stock at present
and wants to end the period with no stock.

	Week 1	Week 2	Week 3	Week 4
No. ordered	4	2	5	7

It costs £60 to set up in a week in which they make
greenhouses. They can make at most 6 greenhouses per week,
at a cost of £150 each. They can store up to two greenhouses
from one week to the next at a cost of £20 each.

Showing your work in tabular form, find the least cost
production schedule for this period.

5 A group of adventurers is trekking across a
desert, starting from S and ending at T.
The trip will take four days. They can
stop at A, B or C at the end of the first,
D or E at the end of the second and F,
G or H at the end of the third.

The table shows the amount of water
(in litres) they would need to carry on
the various legs of the journey. They
would like to devise a route so that their
largest water load is as small as possible.

		To								
		A	B	C	D	E	F	G	H	T
From	S	45	60	59	–	–	–	–	–	–
	A	–	–	–	72	58	–	–	–	–
	B	–	–	–	55	63	–	–	–	–
	C	–	–	–	62	50	–	–	–	–
	D	–	–	–	–	–	50	48	56	–
	E	–	–	–	–	–	63	60	58	–
	F	–	–	–	–	–	–	–	–	46
	G	–	–	–	–	–	–	–	–	62
	H	–	–	–	–	–	–	–	–	57

a Draw a directed network to model
the problem.

b Showing your working on your network,
find the route the group should take.

6 A company plans three projects, A, B and C.
The profit it can make on each project (in £000)
depends on which projects have already been
completed, as shown in the table.

Showing your working in tabular form, use dynamic
programming to find the optimal order in which the
projects should be scheduled and the total profit
which will result.

		Project		
		A	B	C
	None	18	22	25
	A	–	24	28
	B	23	–	26
Already completed	C	24	26	–
	AB	–	–	30
	AC	–	28	–
	BC	25	–	–

7 Exit →

Summary

Refer to

- Bellman's principle of optimality:
    Any part of an optimal path is itself optimal.
- A multi-stage problem consists of a number of states.
    - A state is represented by a vertex of the network.
    - Each state belongs to a stage. A state V belongs to stage n
      if routes from V to the end state have a maximum of n edges.
    - In each state there is a choices of actions, corresponding
      to the edges of the network. Each action has an associated
      weight (cost, profit, distance, etc.).   7.1
- Most dynamic programming problems fall into one of four types:
    - Find the shortest route.
    - Find the longest route.   7.1, 7.2
    - Find the minimax route – the route in which the maximum
      edge weight is as small as possible.
    - Find the maximin route – the route in which the minimum
      edge weight is as large as possible.   7.3

Links

In forestry the long-term profitability of harvesting
timber from a region depends on decisions about
where and when to plant and to fell trees.

The decisions to be made depend on factors such as the
speed of growth of the trees, geographical variations,
the cost of accessing different parts of the region, and so on.

Using dynamic programming allows for these planning
decisions to be made so that profitability is optimised.

D2

1 The diagram shows a network of roads connecting six villages A, B, C, D, E and F. The lengths of the roads are given in km.

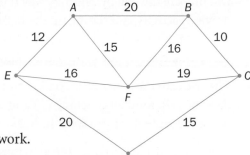

   a Construct a table showing the shortest distances between pairs of villages. You should do this by inspection.

   The table can be taken to represent a complete network.

   b Use the nearest-neighbour algorithm, starting at A, on your table in part **a**. Obtain an upper bound to the length of a tour in this complete network, which starts and finishes at A and visits every village exactly once.

   c Interpret your answer in part **b** in terms of the original network of roads connecting the six villages.

   d By choosing a different vertex as your starting point, use the nearest-neighbour algorithm to obtain a shorter tour than that found in part **b**. State the tour and its length.   [(c) Edexcel Limited 2002]

2 An area manager has to visit branches of his company in seven towns A, B, C, D, E, F and G. The table shows the distances, in km, between these seven towns. The manager lives in town A and plans a route starting and finishing at this town. She wishes to visit each town and drive the minimum distance.

	A	B	C	D	E	F	G
A	–	165	195	280	130	200	150
B	165	–	90	155	150	235	230
C	195	90	–	170	110	175	190
D	280	155	170	–	150	105	163
E	130	150	110	150	–	90	82
F	200	235	175	105	90	–	63
G	150	230	190	163	82	63	–

   a Starting from A, use Prim's algorithm to find a minimum connector and draw the minimum spanning tree. State the order in which you selected the arcs.

   b i Hence determine an initial upper bound for the length of the route planned by the manager.

   ii Starting from your initial upper bound and using a short cut, obtain a route with length less than 870 km.

   iii Find a further cut which produces a route which visits each vertex exactly once and has a length less than 810 km.   [(c) Edexcel Limited 2007]

D2

3 A college wants to offer five full-day activities with a different activity each day from Monday to Friday. The sports hall will only be used for these activities. Each evening the caretaker will prepare the hall by putting away the equipment from the previous activity and setting up the hall for the activity next day. On Friday evening he will put away the equipment used that day and set up the hall for the following Monday.

The five activities offered are badminton (B), cricket nets (C), dancing (D), football coaching (F) and tennis (T). Each will be on the same day from week to week.

The college decides to offer the activities in the order that minimises the total time the caretaker has to spend preparing the hall each week. The hall is initially set up for badminton on Monday.

The table shows the time, in minutes, it will take the caretaker to put away the equipment from one activity and set up the hall for the next.

<table>
<tr><td></td><td></td><td colspan="6" align="center">To</td></tr>
<tr><td></td><td>Time</td><td>B</td><td>C</td><td>D</td><td>F</td><td>T</td></tr>
<tr><td rowspan="5">From</td><td>B</td><td>–</td><td>108</td><td>150</td><td>64</td><td>100</td></tr>
<tr><td>C</td><td>108</td><td>–</td><td>54</td><td>104</td><td>60</td></tr>
<tr><td>D</td><td>150</td><td>54</td><td>–</td><td>150</td><td>102</td></tr>
<tr><td>F</td><td>64</td><td>104</td><td>150</td><td>–</td><td>68</td></tr>
<tr><td>T</td><td>100</td><td>60</td><td>102</td><td>68</td><td>–</td></tr>
</table>

a Explain why this problem is equivalent to the travelling salesman problem.

A possible ordering of activities is

Monday	Tuesday	Wednesday	Thursday	Friday
B	C	D	F	T

b Find the total time taken by the caretaker each week using this ordering.

c Starting with badminton on Monday, use a suitable algorithm to find an ordering that reduces the total time spent each week to less than 7 hours.

d By deleting B, use a suitable algorithm to find a lower bound for the time taken each week. Make your method clear.

[(c) Edexcel Limited 2006]

4 A company wishes to transport its products from three factories F_1, F_2 and F_3 to a single retail outlet R. The capacities of the possible routes, in van loads per day, are shown in the diagram.

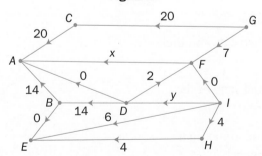

a Copy the diagram and add a supersource, S, to obtain a capacitated network with a single source and a single sink. State the minimum capacity of each arc you have added.

b i State the maximum flow along SF_1ABR and SF_3CR.

  ii Show these maximum flows on your diagram, using numbers in circles.

Taking your answer to part **b ii** as the initial flow pattern,

c i use the labelling procedure to find a maximum flow from S to R. List each flow-augmenting route you find together with its flow.

  ii Prove that your final flow is maximal.

[(c) Edexcel Limited 2002]

5 Figure 1 shows a capacitated directed network. The number on each arc is its capacity.

Figure 1

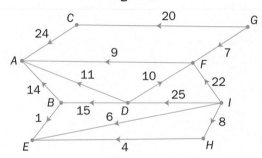

Figure 2 shows a feasible initial flow through the same network.

Figure 2

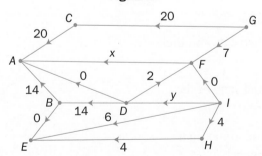

a Write down the values of the flow x and the flow y.

b Obtain the value of the initial flow through the network, and explain how you know it is not maximal.

c Use this initial flow and the labelling procedure to find a maximum flow through the network. You must list each flow-augmenting route you use, together with its flow.

d Show your maximal flow pattern on a copy of the diagram.

e Prove that your flow is maximal.

[(c) Edexcel Limited 2004]

6 The network shows a number of hostels in a national park
and the possible paths joining them. The numbers on the
edges give the lengths, in km, of the paths.

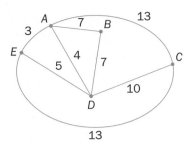

a Draw a complete network showing the shortest distances
between the hostels. (You may do this by inspection.
An algorithm is not required.)

b Use the nearest neighbour algorithm on the complete
network to obtain an upper bound for the length of a
tour in this network which starts at A and visits each
hostel exactly once.

c Interpret your result in part **b** in terms of the
original network.

[(c) Edexcel Limited 2007]

7 The diagram shows a capacitated, directed network.
The capacity of each arc is shown on each arc. The numbers
in circles represent an initial flow from S to T.

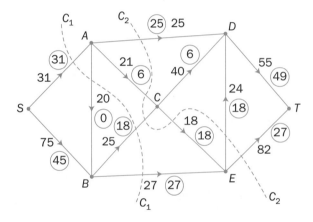

Two cuts C_1 and C_2 are shown on the diagram.

a Write down the capacity of each of the two cuts and the
value of the initial flow.

b On a copy of the diagram set up the initial state of the
labelling procedure.

c Hence use the labelling procedure to find a maximal
flow through the network. You must list each
flow-augmenting path you use, together with its flow.

d Show your maximal flow pattern on a copy of the diagram.

e Prove that your flow is maximal.

[(c) Edexcel Limited 2006]

8 Kris produces custom made racing cycles. She can produce up to four cycles each month, but if she wishes to produce more than three in any one month she has to hire additional help at a cost of £350 for that month. In any month when cycles are produced, the overhead costs are £200. A maximum of three cycles can be held in stock in any one month, at a cost of £40 per cycle per month. Cycles must be delivered at the end of the month. The order book for cycles is

Month	August	September	October	November
Number of cycles required	3	3	5	2

Disregarding the cost of parts and Kris's time

a determine the total cost of storing two cycles and producing four cycles in a given month, making your calculations clear.

There is no stock at the beginning of August and Kris plans to have no stock after the November delivery.

b Use dynamic programming to determine the production schedule which minimises the costs, showing your working in a table.

The fixed cost of parts is £600 per cycle and of Kris's time is £500 per month. She sells the cycles for £2000 each.

c Determine her total profit for the four-month period.

[(c) Edexcel Limited 2003]

9 Jenny wishes to travel from S to T. There are several routes available. She wishes to choose the route on which the maximum altitude, above sea level, is as small as possible. This is called the minimax route.

The diagram shows the possible routes and the weights on the edges give the maximum altitude on the road (in units of 100 feet).

Use dynamic programming to determine the route or routes Jenny should take. Show your calculations in a table with columns labelled as shown.

Stage	Initial State	Action	Final State	Value

[(c) Edexcel Limited 2007]

10 Joan sells ice cream. She needs to decide which three shows to visit over a three-week period in the summer. She starts the three-week period at home and finishes at home. She will spend one week at each of the three shows she chooses travelling directly from one show to the next.

Table 1 gives the week in which each show is held.
Table 2 gives the expected profits from visiting each show.
Table 3 gives the cost of travel between shows.

Table 1

Week	1	2	3
Shows	A, B, C	D, E	F, G, H

Table 2

Show	A	B	C	D	E	F	G	H
Expected Profit (£)	900	800	1000	1500	1300	500	700	600

Table 3

Travel costs (£)	A	B	C	D	E	F	G	H
Home	70	80	150			80	90	70
A				180	150			
B				140	120			
C				200	210			
D						200	160	120
E						170	100	110

It is decided to use dynamic programming to find a schedule that maximises the total expected profit, taking into account the travel costs.

a Define suitable stage, state and action variables.

b Determine the schedule that maximises the total profit. Show your working in a table.

c Advise Joan on the shows that she should visit and state her total expected profit.

[(c) Edexcel Limited 2004]

Answers

Chapter 1

Check in

1
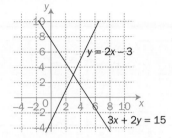

2
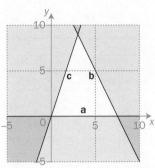

3   $x = -1, y = 2, z = 3$

Exercise 1.1

1   Make x kg of Assolato, y kg of Buona Salute, z kg of Contadino. Maximise $P = 0.6x + 0.5y + 0.9z$, subject to $3x + 2y + 3z \leqslant 5000$, $2x + 3y + 4z \leqslant 6000$, $x + y + 2z \leqslant 4000$, $x \geqslant 0$, $y \geqslant 0$, $z \geqslant 0$

2   x ml of glucose, y ml of oil, z ml of water. Minimise $C = 1.4x + 0.9y + 0.2z$, subject to $x + y + z = 1000$, $z \leqslant 600$, $y \leqslant 2x$, $x \leqslant 3y$, $x \geqslant 0$, $y \geqslant 0$, $z \geqslant 0$. This can be simplified to: x ml of glucose, y ml of oil Minimise $C = 1.2x + 0.7y + 200$, subject to $x + y \geqslant 400$, $y \leqslant 2x$, $x \leqslant 3y$, $x \geqslant 0$, $y \geqslant 0$,

3   w of type A, x of B, y of C, z of D. Maximise $P = 50w + 30x + 80y + 60z$, subject to $6w + 4x + 8y + 5z \leqslant 800$, $2w + x + 2y + z \leqslant 240$, $w + x + 2y + 2z \leqslant 300$, $w + 2x + 2y + z \leqslant 320$, $w, x, y, z \geqslant 0$, w, x, y and z are all integers.

Exercise 1.2

1   a
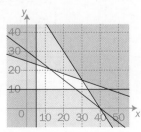

$P = 150$ when $x = 30, y = 15$

b
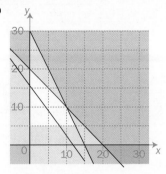

$P = 70$ when $x = 10, y = 10$

c
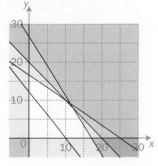

$C = 49$ when $x = 6, y = 5$

d
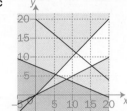

$P = 95.6$ when $x = 12.4, y = 8.4$

2
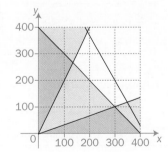

$C = 546\frac{2}{3}$ when $x = 133\frac{1}{3}, y = 266\frac{2}{3}, z = 600$

3   Max profit = £1106.25 with 2000 litres Frute-kup and 2750 litres Fresh-squeeze

Exercise 1.3

1 a

Basic variable	x	y	s	t	Value	Row no.
s	1	2	1	0	40	[1]
t	3	2	0	1	60	[2]
P	–1	–1	0	0	0	[3]

s	0	$1\frac{1}{3}$	1	$-\frac{1}{3}$	20	$[4]=[1]-[5]$
x	1	$\frac{2}{3}$	0	$\frac{1}{3}$	20	$[5]=[2]\div 3$
P	0	$-\frac{1}{3}$	0	$\frac{1}{3}$	20	$[6]=[3]+[5]$

y	0	1	$\frac{3}{4}$	$-\frac{1}{4}$	15	$[7]=[4]\div 1\frac{1}{3}$
x	1	0	$-\frac{1}{2}$	$\frac{1}{2}$	10	$[8]=[5]-[7]\times\frac{2}{3}$
P	0	0	$\frac{1}{4}$	$\frac{1}{4}$	25	$[9]=[6]+[7]\div 3$

Maximum $P = 25$ when $x = 10$, $y = 15$

b

Basic variable	x	y	s	t	Value	Row no.
s	4	3	1	0	170	[1]
t	5	2	0	1	160	[2]
P	–2	–1	0	0	0	[3]

s	0	$1\frac{2}{5}$	1	$-\frac{4}{5}$	42	$[4]=[1]-[5]\times 4$
x	1	$\frac{2}{5}$	0	$\frac{1}{5}$	32	$[5]=[2]\div 5$
P	0	$-\frac{1}{5}$	0	$\frac{2}{5}$	64	$[6]=[3]+[5]\times 2$

y	0	1	$\frac{5}{7}$	$-\frac{4}{7}$	30	$[7]=[4]\div 1\frac{2}{5}$
x	1	0	$-\frac{2}{7}$	$\frac{3}{7}$	20	$[8]=[5]-[7]\times\frac{2}{5}$
P	0	0	$\frac{1}{7}$	$\frac{2}{7}$	70	$[9]=[6]+\frac{[7]}{5}$

Maximum $P = 70$ when $x = 20$, $y = 30$

c

Basic variable	x	y	s	t	u	Value	Row no.
s	2	3	1	0	0	30	[1]
t	1	3	0	1	0	24	[2]
u	4	3	0	0	1	48	[3]
P	–4	–5	0	0	0	0	[4]

Basic variable	x	y	s	t	u	Value	Row no.
s	1	0	1	–1	0	6	$[5]=[1]-3\times[6]$
y	$\frac{1}{3}$	1	0	$\frac{1}{3}$	0	8	$[6]=\frac{[2]}{3}$
u	3	0	0	–1	1	24	$[7]=[3]-[6]\times 3$
P	$-2\frac{1}{3}$	0	0	$1\frac{2}{3}$	0	40	$[8]=[4]+5\times[6]$

Basic variable	x	y	s	t	u	Value	Row no.
x	1	0	1	–1	0	6	$[9]=[5]$
y	0	1	$-\frac{1}{3}$	$\frac{2}{3}$	0	6	$[10]=[6]-\frac{[9]}{3}$
u	0	0	–3	2	1	6	$[11]=[7]-3\times[9]$
P	0	0	$2\frac{1}{3}$	$-\frac{2}{3}$	0	54	$[12]=[8]+[9]\times\frac{7}{3}$

Basic variable	x	y	s	t	u	Value	Row no.
x	1	0	$-\frac{1}{2}$	0	$\frac{1}{2}$	9	$[13]=[9]+[15]$
y	0	1	$\frac{2}{3}$	0	$-\frac{1}{3}$	4	$[14]=[10]-[15]\times\frac{2}{3}$
t	0	0	$-\frac{3}{2}$	1	$\frac{1}{2}$	3	$[15]=\frac{[11]}{2}$
P	0	0	$1\frac{1}{3}$	0	$\frac{1}{3}$	56	$[16]=[12]+[15]\times\frac{2}{3}$

Maximum $P = 56$ when $x = 9$, $y = 4$ $(t = 3)$

d

Basic variable	x	y	s	t	u	Value	Row no.
s	1	1	1	0	0	6	[1]
t	2	1	0	1	0	9	[2]
u	3	2	0	0	1	15	[3]
P	–1	–2	0	0	0	0	[4]

Basic variable	x	y	s	t	u	Value	Row no.
y	1	1	1	0	0	6	$[5]=[1]$
t	1	0	–1	1	0	3	$[6]=[2]-[5]$
u	1	0	–2	0	1	3	$[7]=[3]-[5]\times 2$
P	1	0	2	0	0	12	$[8]=[4]+[5]\times 2$

Maximum $P = 12$ when $x = 0$, $y = 6$ $(t = 3, u = 3)$

e

Basic variable	x	y	z	s	t	Value	Row no.
s	8	5	2	1	0	7	[1]
t	1	2	3	0	1	4	[2]
P	–3	–4	–2	0	0	0	[3]

Basic variable	x	y	z	s	t	Value	Row no.
y	1.6	1	0.4	0.2	0	1.4	$[4] = \frac{[1]}{5}$
t	-2.2	0	2.2	-0.4	1	1.2	$[5] = [2] - 2 \times [4]$
P	3.4	0	-0.4	0.8	0	5.6	$[6] = [3] + 4 \times [4]$

Basic variable	x	y	z	s	t	Value	Row no.
y	2	1	0	$\frac{3}{11}$	$-\frac{2}{11}$	$1\frac{2}{11}$	$[7] = [4] - [8] \times 0.4$
z	-1	0	1	$-\frac{2}{11}$	$\frac{5}{11}$	$\frac{6}{11}$	$[8] = \frac{[5]}{2.2}$
P	3	0	0	$\frac{7}{11}$	$\frac{2}{11}$	$5\frac{9}{11}$	$[9] = [6] + [8] \times 0.4$

Maximum $P = 5\frac{9}{11}$ when $x = 0$, $y = 1\frac{2}{11}$, $z = \frac{6}{11}$

f

Basic variable	x	y	z	s	t	u	Value	Row no.
s	2	4	-1	1	0	0	7	[1]
t	5	6	2	0	1	0	16	[2]
u	7	7	4	0	0	1	25	[3]
P	-7	-6	-4	0	0	0	0	[4]

Basic variable	x	y	z	s	t	u	Value	Row no.
s	0	1.6	-1.8	1	-0.4	0	0.6	$[5] = [1] - [6] \times 2$
x	1	1.2	0.4	0	0.2	0	3.2	$[6] = \frac{[2]}{5}$
u	0	-1.4	1.2	0	-1.4	1	2.6	$[7] = [3] - [6] \times 7$
P	0	2.4	-1.2	0	1.4	0	22.4	$[8] = [4] + [6] \times 7$

Basic variable	x	y	z	s	t	u	Value	Row no.
s	0	-0.5	0	1	-2.5	1.5	4.5	$[9] = [5] + [11] \times 1.8$
x	1	$\frac{5}{3}$	0	0	$\frac{2}{3}$	$-\frac{1}{3}$	$2\frac{1}{3}$	$[10] = [6] - [11] \times 0.4$
z	0	$-\frac{7}{6}$	1	0	$-\frac{7}{6}$	$\frac{5}{6}$	$2\frac{1}{6}$	$[11] = \frac{[7]}{1.2}$
P	0	1	0	0	0	1	25	$[12] = [8] + [11] \times 1.2$

Maximum $P = 25$ when $x = 2\frac{1}{3}$, $y = 0$, $z = 2\frac{1}{6}$ $\left(s = 4\frac{1}{2}\right)$

2 a Make x kg Regular, y kg Luxury. Maximise
$P = 1.7x + 2.5y$ subject to $0.7x + 0.5y + s = 8000$,
$0.3x + 0.5y + t = 6000$.

b

Basic variable	x	y	s	t	Value	Row no.
s	0.7	0.5	1	0	8000	[1]
t	0.3	0.5	0	1	6000	[2]
P	-1.7	-2.5	0	0	0	[3]

	x	y	s	t	Value	Row no.
s	0.4	0	1	-1	2000	$[4] = [1] - \frac{[5]}{2}$
y	0.6	1	0	2	12000	$[5] = [2] \times 2$
P	-0.2	0	0	5	30000	$[6] = [3] + [5] \times 2.5$

	x	y	s	t	Value	Row no.
x	1	0	$\frac{5}{2}$	$-\frac{5}{2}$	5000	$[7] = [4] \times 2.5$
y	0	1	$-\frac{3}{2}$	$-\frac{7}{2}$	9000	$[8] = [5] - [7] \times 0.6$
P	0	0	$\frac{1}{2}$	$\frac{9}{2}$	31000	$[9] = [6] + [7] \times 0.2$

$P = £31\,000$, $x = 5000$, $y = 9000$

3 a $P = 3x + y + 2z$, $2x + 3y + z \leqslant 20$, $x + 2y + z \leqslant 12$, $x \geqslant 0$, $y \geqslant 0$, $z \geqslant 0$

b

Basic variable	x	y	z	s	t	Value	Row no.
s	2	3	1	1	0	20	[1]
t	1	2	1	0	1	12	[2]
P	-3	-1	-2	0	0	0	[3]

Basic variable	x	y	z	s	t	Value	Row no.
x	1	$\frac{3}{2}$	$\frac{1}{2}$	$\frac{1}{2}$	0	10	$[4] = \frac{[1]}{2}$
t	0	$\frac{1}{2}$	$\frac{1}{2}$	$-\frac{1}{2}$	1	2	$[5] = [2] - [4]$
P	0	$\frac{7}{2}$	$-\frac{1}{2}$	$\frac{3}{2}$	0	30	$[6] = [3] + 3 \times [4]$

Not optimal as objective row contains a negative entry. $x = 10$, $y = 0$, $z = 0$, $s = 0$, $t = 2$

c

Basic variable	x	y	z	s	t	Value	Row no.
x	1	1	0	1	-1	8	$[7] = [4] - \frac{[8]}{2}$
z	0	1	1	-1	2	4	$[8] = [5] \times 2$
P	0	4	0	1	1	32	$[9] = [6] + \frac{[8]}{2}$

$P = 32$, $x = 8$, $y = 0$, $z = 4$, $s = 0$, $t = 0$

4 a $P - x - 3y = 0$, $x + y + s = 6$, $x + 4y + t = 12$,
$x + 2y + u = 7$

b

Basic variable	x	y	s	t	u	Value	Row no.
s	1	1	1	0	0	6	[1]
t	1	4	0	1	0	12	[2]
u	1	2	0	0	1	7	[3]
P	-1	-3	0	0	0	0	[4]

Basic variable	x	y	s	t	u	Value	Row no.
s	$\frac{3}{4}$	0	1	$-\frac{1}{4}$	0	3	$[5]=[1]-[6]$
y	$\frac{1}{4}$	1	0	$\frac{1}{4}$	0	3	$[6]=\frac{[2]}{4}$
u	$\frac{1}{2}$	0	0	$-\frac{1}{2}$	1	1	$[7]=[3]-[6]\times 2$
P	$-\frac{1}{4}$	0	0	$\frac{3}{4}$	0	9	$[8]=[4]+[6]\times 3$

Negative entry in objective row.

c

Basic variable	x	y	s	t	u	Value	Row no.
s	0	0	1	$\frac{1}{4}$	$-\frac{3}{2}$	$\frac{3}{2}$	$[9]=[5]-[11]\times\frac{3}{4}$
y	0	1	0	$\frac{1}{2}$	$-\frac{1}{2}$	$\frac{5}{2}$	$[10]=\frac{[6]}{4}$
x	1	0	0	-1	2	2	$[11]=[7]\times 2$
P	0	0	0	$\frac{1}{2}$	$\frac{1}{2}$	$\frac{19}{2}$	$[12]=[8]+\frac{[1]}{4}$

$P = 9.5$, $x = 2$, $y = 2.5$ $(s = 1.5)$

5

Basic variable	x	y	z	s	t	u	Value	Row no.
s	1	1	2	1	0	0	40	[1]
t	2	3	0	0	1	0	20	[2]
u	1	2	2	0	0	1	30	[3]
P	-3	-2	-1	0	0	0	0	[4]

Basic variable	x	y	z	s	t	u	Value	Row no.
s	0	$-\frac{1}{2}$	2	1	$-\frac{1}{2}$	0	30	$[5]=[1]-[6]$
x	1	$\frac{3}{2}$	0	0	$\frac{1}{2}$	0	10	$[6]=\frac{[2]}{2}$
u	0	$\frac{1}{2}$	2	0	$-\frac{1}{2}$	1	20	$[7]=[3]-[6]$
P	0	$\frac{5}{2}$	-1	0	$\frac{3}{2}$	0	30	$[8]=[4]+[6]\times 3$

Basic variable	x	y	z	s	t	u	Value	Row no.
s	0	-1	0	1	0	-1	10	$[9]=[5]+[11]\times 2$
x	1	$\frac{3}{2}$	0	0	$\frac{1}{2}$	0	10	$[10]=[6]$
z	0	$\frac{1}{4}$	1	0	$-\frac{1}{4}$	$\frac{1}{2}$	10	$[11]=\frac{[7]}{2}$
P	0	$\frac{11}{4}$	0	0	$\frac{5}{4}$	$\frac{1}{2}$	40	$[12]=[8]+[11]$

$P = 40$, $x = 10$, $y = 0$, $z = 10$ $(s = 10)$

6 a Maximise $P = 5x + 6y$ subject to $3x + 3y \leqslant 40$, $x + 2y \leqslant 25$

b

Basic variable	x	y	s	t	Value	Row no.
s	3	3	1	0	40	[1]
t	1	2	0	1	25	[2]
P	-5	-6	0	0	0	[3]

Basic variable	x	y	s	t	Value	Row no.
s	$\frac{3}{2}$	0	1	$-\frac{3}{2}$	$\frac{5}{2}$	$[4]=[1]-[5]\times 3$
y	$\frac{1}{2}$	1	0	$\frac{1}{2}$	$\frac{25}{2}$	$[5]=\frac{[2]}{2}$
P	-2	0	0	3	75	$[6]=[3]+[5]\times 6$

$P = 75$, $x = 0$, $y = 12.5$, $s = 2.5$, $t = 0$. Negative entry in objective row.

c

Basic variable	x	y	s	t	Value	Row no.
x	1	0	$\frac{2}{3}$	-1	$1\frac{2}{3}$	$[7]=[4]\times\frac{2}{3}$
y	0	1	$-\frac{1}{3}$	1	$11\frac{2}{3}$	$[8]=[5]-\frac{[7]}{2}$
P	0	0	$\frac{4}{3}$	1	$78\frac{1}{3}$	$[9]=[6]+[7]\times 2$

$P = 78\frac{1}{3}$, $x = 1\frac{2}{3}$, $y = 11\frac{2}{3}$

7 b

Basic variable	x	y	s	t	Value	Row no.
s	3	2	1	0	14 400	[1]
t	2	3	0	1	15 000	[2]
P	-1	-1	0	0	0	[3]

	x	y	s	t	Value	Row no.
x	1	$\frac{2}{3}$	$\frac{1}{3}$	0	4800	$[4]=\frac{[1]}{3}$
t	0	$\frac{5}{3}$	$-\frac{2}{3}$	1	5400	$[5]=[2]-[4]\times 2$
P	0	$-\frac{1}{3}$	$\frac{1}{3}$	0	4800	$[6]=[3]+[4]$

	x	y	s	t	Value	Row no.
x	1	0	$\frac{3}{5}$	$-\frac{9}{25}$	2640	$[7]=[4]-[8]\times\frac{2}{3}$
y	0	1	$-\frac{2}{5}$	$\frac{3}{5}$	3240	$[8]=[5]\times\frac{3}{5}$
P	0	0	$\frac{1}{5}$	$\frac{1}{5}$	5880	$[9]=[6]+\frac{[8]}{3}$

$D = 5880\,\text{m}$, $x = 2640$, $y = 3240$

8

Basic variable	x	y	z	s	t	u	Value	Row no.
s	3	2	3	1	0	0	5000	[1]
t	2	3	4	0	1	0	6000	[2]
u	1	1	2	0	0	1	4000	[3]
P	-0.6	-0.5	-0.9	0	0	0	0	[4]

Basic variable	x	y	z	s	t	u	Value	Row no.
s	$\frac{3}{2}$	$-\frac{1}{4}$	0	1	$-\frac{3}{4}$	0	500	$[5]=[1]-[6]\times 3$
z	$\frac{1}{2}$	$\frac{3}{4}$	1	0	$\frac{1}{4}$	0	1500	$[6]=\frac{[2]}{4}$
u	0	$-\frac{1}{2}$	0	0	$-\frac{1}{2}$	1	1000	$[7]=[3]-[6]\times 2$
P	$-\frac{3}{20}$	$\frac{7}{40}$	0	0	$\frac{9}{40}$	0	1350	$[8]=[4]+[6]\times 0.9$

Basic variable	x	y	z	s	t	u	Value	Row no.
x	1	$-\frac{1}{6}$	0	$\frac{2}{3}$	$-\frac{1}{2}$	0	$333\frac{1}{3}$	$[9]=[5]\times\frac{2}{3}$
z	0	$\frac{5}{6}$	1	$-\frac{1}{2}$	$\frac{1}{2}$	0	$1333\frac{1}{3}$	$[10]=[6]-\frac{[9]}{2}$
u	0	$-\frac{1}{2}$	0	0	$-\frac{1}{2}$	1	1000	$[11]=[7]$
P	0	$\frac{3}{20}$	0	$\frac{1}{10}$	$\frac{3}{20}$	0	1400	$[12]=[8]+[9]\times\frac{3}{20}$

$P = 1400$, $x = 333\frac{1}{3}$, $y = 0$, $z = 1333\frac{1}{3}$ $(u = 1000)$

9 a

Basic variable	w	x	y	z	r	s	t	u	Value
r	6	4	8	5	1	0	0	0	800
s	2	1	2	1	0	1	0	0	240
t	1	1	2	2	0	0	1	0	300
u	1	2	2	1	0	0	0	1	320
P	-50	-30	-80	-60	0	0	0	0	0

b

y	$\frac{3}{4}$	0.5	1	$\frac{5}{8}$	$\frac{1}{8}$	0	0	0	100
s	$\frac{1}{2}$	0	0	$-\frac{1}{4}$	$-\frac{1}{4}$	1	0	0	40
t	$-\frac{1}{2}$	0	0	$\frac{3}{4}$	$-\frac{1}{4}$	0	1	0	100
u	$\frac{1}{2}$	1	0	$-\frac{1}{4}$	$-\frac{1}{4}$	0	0	1	120
P	10	10	0	-10	10	0	0	0	8000

y	$\frac{7}{6}$	0.5	1	0	$\frac{1}{3}$	0	-0.8	0	$16\frac{2}{3}$
s	$\frac{1}{3}$	0	0	0	-1.3	1	$\frac{1}{3}$	0	$73\frac{1}{3}$
z	$-\frac{2}{3}$	0	0	1	$-\frac{1}{3}$	0	$\frac{4}{3}$	0	$133\frac{1}{3}$
u	$-\frac{2}{3}$	1	0	0	$-\frac{1}{3}$	0	$\frac{1}{3}$	1	$153\frac{1}{3}$
P	$3\frac{1}{3}$	10	0	0	$6\frac{2}{3}$	0	13.3	0	$9333\frac{1}{3}$

c This is only an approximate answer, as the variables should be integers. Trial and error suggests that $y = 16$, $z = 134$, giving $P = 9320$, is the best result.

10 a $x + 2y - s = 20$, $3x + y - t = 30$. Negative coefficients for the slack variables.

b $y = 0$, $s = 0$, $x = 20$, $t = 30$

c

Basic variable	x	y	s	t	Value	Row no.
x	1	2	-1	0	20	[1]
t	0	5	-3	1	30	[2]
(−C)	0	-1	1	0	-20	[3]

x	1	0	$\frac{1}{5}$	$-\frac{2}{5}$	8	$[4]=[1]-[5]\times 2$
y	0	1	$-\frac{3}{5}$	$\frac{1}{5}$	6	$[5]=\frac{[2]}{5}$
P	0	0	$\frac{2}{5}$	$\frac{1}{5}$	-14	$[6]=[3]+[5]$

d $C = 14$ when $x = 8$, $y = 6$

Review 1

1 a Maximise $P = 8x + 5y + 6z$ subject to
$2x + y + 2z + s = 400$, $x + 3y + z + t = 300$,
$2x + y + z + u = 500$

b

Basic variable	x	y	z	s	t	u	Value	Row no.
s	2	1	2	1	0	0	400	[1]
t	1	3	1	0	1	0	300	[2]
u	2	1	1	0	0	1	500	[3]
P	-8	-5	-6	0	0	0	0	[4]

Basic variable	x	y	z	s	t	u	Value	Row no.
x	1	$\frac{1}{2}$	1	$\frac{1}{2}$	0	0	200	$[5]=\frac{[1]}{2}$
t	0	$\frac{5}{2}$	0	$-\frac{1}{2}$	1	0	100	$[6]=[2]-[5]$
u	0	0	-1	-1	0	1	100	$[7]=[3]-[5]\times 2$
P	0	-1	2	4	0	0	1600	$[8]=[4]+[5]\times 8$

The objective row has a negative entry.

c i

Basic variable	x	y	z	s	t	u	Value	Row no.
x	1	0	1	$\frac{3}{5}$	$-\frac{1}{5}$	0	180	$[9]=[5]-\frac{[10]}{2}$
y	0	1	0	$-\frac{1}{5}$	$\frac{2}{5}$	0	40	$[10]=[6]\times\frac{2}{5}$
u	0	0	-1	-1	0	1	100	$[11]=[7]$
P	0	0	2	$\frac{19}{5}$	$\frac{2}{5}$	0	1640	$[12]=[8]+[10]$

ii Profit £1640. Make 180 type A, 40 type B, no type C. 100 packets of purple bulbs left.

2 a £400

b $5x + 6y + 3z \leqslant 90$, $2x + 4y + z \leqslant 42$

c

Basic variable	x	y	z	s	t	Value	Row no.
s	5	6	3	1	0	90	[1]
t	2	4	1	0	1	42	[2]
P	-4	-5	-2	0	0	0	[3]

D2

Basic variable	x	y	z	s	t	Value	Row no.
s	2	0	$\frac{3}{2}$	1	$-\frac{3}{2}$	27	$[4] = [1] - [5] \times 6$
y	$\frac{1}{2}$	1	$\frac{1}{4}$	0	$\frac{1}{4}$	$10\frac{1}{2}$	$[5] = \frac{[2]}{4}$
P	$-\frac{3}{2}$	0	$-\frac{3}{4}$	0	$\frac{5}{4}$	$52\frac{1}{2}$	$[6] = [3] + [5] \times 5$

Basic variable	x	y	z	s	t	Value	Row no.
x	1	0	$\frac{3}{4}$	$\frac{1}{2}$	$-\frac{3}{4}$	$13\frac{1}{2}$	$[7] = \frac{[4]}{2}$
y	0	1	$-\frac{1}{8}$	$-\frac{1}{4}$	$\frac{5}{8}$	$3\frac{3}{4}$	$[8] = [5] - \frac{[7]}{2}$
P	0	0	$\frac{3}{8}$	$\frac{3}{4}$	$\frac{1}{8}$	$72\frac{3}{4}$	$[9] = [6] + [7] \times \frac{3}{2}$

No negative entries left in the objective row.

d Cannot make fractions of a bicycle.

e 13 type A, 4 type B, no type C.

3 a Maximise $P = 50x + 60y + 40z$ subject to
$x + y + 2z + s = 100$, $2x + y + 2z + t = 120$,
$x + 2y + 2z + u = 80$

b

Basic variable	x	y	z	s	t	u	Value	Row no.
s	1	1	2	1	0	0	100	[1]
t	2	1	3	0	1	0	120	[2]
u	1	2	2	0	0	1	80	[3]
P	-50	-60	-40	0	0	0	0	[4]

Basic variable	x	y	z	s	t	u	Value	Row no.
s	$\frac{1}{2}$	0	1	1	0	$-\frac{1}{2}$	60	$[5] = [1] - [6] \times 3$
t	$\frac{3}{2}$	0	2	0	1	$-\frac{1}{2}$	80	$[6] = \frac{[2]}{4}$
y	$\frac{1}{2}$	1	1	0	0	$\frac{1}{2}$	40	$[7] = [3] - [6] \times 2$
P	-20	0	20	0	0	30	2400	$[8] = [4] + [6] \times 0.9$

Basic variable	x	y	z	s	t	u	Value	Row no.
x	0	0	$\frac{1}{3}$	1	$-\frac{1}{2}$	$-\frac{1}{2}$	$33\frac{1}{3}$	$[9] = [5] \times \frac{2}{3}$
z	1	0	$\frac{4}{3}$	0	$\frac{2}{3}$	$-\frac{1}{3}$	$53\frac{1}{3}$	$[10] = [6] - \frac{[9]}{2}$
u	0	1	$\frac{1}{3}$	0	$-\frac{1}{3}$	$\frac{2}{3}$	$13\frac{1}{3}$	$[11] = [7]$
P	0	0	$46\frac{2}{3}$	0	$13\frac{1}{3}$	$23\frac{1}{3}$	$3466\frac{2}{3}$	$[12] = [8] + [9] \times \frac{3}{20}$

c Cannot make fractions of bed frame, so try integers near the answer found. Best is 52 type A, 14 type B, no type C, profit £3440.

4 a No negative entries in objective row.

b $x = 1$, $y = 0$, $z = 0.5$, $s = 2$, $t = 0$, $u = 0$, $P = 15$

c $P = 15 - 2y - 2t - u$. If y, t or u increase, P will be below 15.

Chapter 2
Exercise 2.1

1 a

From \ To	X	Y	Z	Supply
A	200	–	–	200
B	10	130	–	140
C	–	110	120	230
Demand	210	240	120	570

b

From \ To	X	Y	Z	Supply
A	80	20	–	100
B	–	130	–	130
C	–	–	90	90
Demand	80	150	90	320

c

From \ To	P	Q	R	S	Supply
A	850	110	–	–	960
B	–	480	270	–	750
C	–	–	400	420	820
Demand	850	590	670	420	2530

d

From \ To	X	Y	Z	Supply
A	60	–	–	60
B	80	–	–	80
C	10	90	10	110
D	–	–	120	120
Demand	150	90	130	370

2 a Total supply = total demand = 360, so problem is balanced.

b

From \ To	X	Y	Z	Supply (Availability)
A	[2] 85	[3] 35	[2] –	120
B	[4] –	[6] 95	[3] –	95
C	[2] –	[4] 20	[4] 125	145
Demand (Requirement)	85	150	125	360

c Cost = £1425

Exercise 2.2

1 a

R \ K	$K_1 = 6$	$K_2 = 4$	$K_3 = 2$
$R_1 = 0$	[6] X	[4] X	[5] ③
$R_2 = 4$	[7] (−3)	[8] X	[8] ②
$R_3 = 1$	[7] ⓪	[5] X	[3] X

Allocation is not optimal.

b

R \ K	$K_1 = 3$	$K_2 = 2$	$K_3 = 4$
$R_1 = 0$	[3] X	[3] ①	[4] ⓪
$R_2 = 2$	[7] ②	[4] X	[6] X
$R_3 = 1$	[4] X	[5] ②	[5] X

Allocation is optimal.

c

R \ K	$K_1 = 6$	$K_2 = 8$	$K_3 = 6$
$R_1 = 0$	[6] X	[9] ①	[6] ⓪
$R_2 = -5$	[3] ②	[3] X	[2] ①
$R_3 = -4$	[2] X	[3] (−1)	[2] X
$R_4 = -3$	[3] ⓪	[5] X	[3] X

Allocation is not optimal.

d

R \ K	$K_1 = 4$	$K_2 = 7$	$K_3 = 3$	$K_4 = 8$
$R_1 = 0$	[5] ①	[7] ⓪	[4] ①	[8] X
$R_2 = 1$	[7] ②	[9] ①	[4] X	[9] X
$R_3 = 0$	[4] ⓪	[7] X	[3] X	[9] ①
$R_4 = -2$	[2] X	[5] X	[2] ①	[7] ①

Allocation is optimal.

2

From \ To	X	Y	Z	Supply
A	[4] 70	[3] 20	[4] –	90
B	[7] –	[5] 60	[6] 70	130
C	[6] –	[4] –	[5] 80	80
Demand	70	80	150	300

R \ K	$K_1 = 5$	$K_2 = 9$	$K_3 = 8$
$R_1 = 0$	[5] X	[9] X	[6] ⓪
$R_2 = -3$	[3] ①	[6] X	[5] X
$R_3 = -5$	[4] ①	[7] ⓪	[3] X

Allocation is optimal.

3 a, b

From \ To	X	Y	Z	Supply
A	[14] 18	[10] –	[16] –	18
B	[18] 2	[15] 10	[8] –	12
C	[9] –	[11] 6	[17] 10	16
Demand	20	16	10	46

c Total distance = 674 km

d

R \ K	$K_1 = 14$	$K_2 = 11$	$K_3 = 17$
$R_1 = 0$	[14] X	[10] (−1)	[16] (−1)
$R_2 = 4$	[18] X	[15] X	[8] (−13)
$R_3 = 0$	[9] (−5)	[11] X	[17] X

Allocation is not optimal.

4 a

From \ To	X	Y	Z	Supply
A	[7] 80	[7] 20	[1] –	100
B	[2] –	[6] 160	[5] –	160
C	[3] –	[4] 10	[6] 110	120
Demand	80	190	110	380

R \ K	$K_1 = 7$	$K_2 = 7$	$K_3 = 9$
$R_1 = 0$	[7] X	[7] X	[1] (−8)
$R_2 = -1$	[2] (−4)	[6] X	[5] (−3)
$R_3 = -3$	[3] (−1)	[4] X	[6] X

Not optimal (negative I values)

D2

b

From \ To	X	Y	Z	Supply
A	[7] –	[7] –	[1] 100	100
B	[2] –	[6] 150	[5] 10	160
C	[3] 80	[4] 40	[6] –	120
Demand	80	190	110	380

	$K_1 = 1$	$K_2 = 2$	$K_3 = 1$
$R_1 = 0$	[7] ⑥	[7] ⑤	[1] X
$R_2 = 4$	[2] ⑨⁻³	[6] X	[5] X
$R_3 = 2$	[3] X	[4] X	[6] ③

Not optimal (negative I values)

c

From \ To	X	Y	Z	Supply
A	[7] –	[7] –	[1] 100	100
B	[2] 80	[6] 70	[5] 10	160
C	[3] –	[4] 120	[6] –	120
Demand	80	190	110	380

	$K_1 = -2$	$K_2 = 2$	$K_3 = 1$
$R_1 = 0$	[7] ⑨	[7] ⑤	[1] X
$R_2 = 4$	[2] X	[6] X	[5] X
$R_3 = 2$	[3] ③	[4] X	[6] ③

Optimal (no negative I values)

Exercise 2.3

1 a

50		50
110	80	
	140	

Entering cell $(1,3)$, exiting cell $(2,3)$

b

170	150	
		90
	190	140

Entering cell $(2,3)$, exiting cell $(2,2)$

c

440			260
660			
	500		
	480	150	40

Entering cell $(1,4)$, exiting cell $(2,2)$

2 a

From \ To	X	Y	Z	Supply
A	[8] 230	[6] –	[8] –	230
B	[6] 70	[3] 210	[4] 70	350
C	[3] –	[4] –	[3] 200	200
Demand	300	210	270	780

Total cost = £3770

b

	$K_1 = 8$	$K_2 = 5$	$K_3 = 6$
$R_1 = 0$	[8] X	[6] ①	[8] ②
$R_2 = -2$	[6] X	[3] X	[4] X
$R_3 = -3$	[3] ⁻②	[4] ②	[3] X

Improve by using cell $(3,1)$

From \ To	X	Y	Z	Supply
A	[8] 230	[6] –	[8] –	230
B	[6] –	[3] 210	[4] 140	350
C	[3] 70	[4] –	[3] 130	200
Demand	300	210	270	780

Entering cell $(3,1)$
Exiting cell $(2,1)$

	$K_1 = 8$	$K_2 = 7$	$K_3 = 8$
$R_1 = 0$	[8] X	[6] ⁻①	[8] ⓪
$R_2 = -4$	[6] ②	[3] X	[4] X
$R_3 = -5$	[3] X	[4] ②	[3] X

Not yet optimal

From \ To	X	Y	Z	Supply
A	[8] 100	[6] 130	[8] –	230
B	[6] –	[3] 80	[4] 270	350
C	[3] 200	[4] –	[3] –	200
Demand	300	210	270	780

Entering cell $(1,2)$
Exiting cell $(3,3)$

D2

	$K_1 = 8$	$K_2 = 6$	$K_3 = 7$
$R_1 = 0$	8 X	6 X	8 ①
$R_2 = -3$	6 ①	3 X	4 X
$R_3 = -5$	3 X	4 ③	3 ①

Allocation is optimal.
Total cost = £3500

3 a

From \ To	P	Q	R	S	Supply
A	4 70	2 10	7 –	4 –	80
B	5 –	2 30	5 40	3 40	110
C	2 –	4 –	5 –	2 60	60
Demand	70	40	40	100	250

b

	$K_1 = 4$	$K_2 = 2$	$K_3 = 5$	$K_4 = 3$
$R_1 = 0$	4 X	2 X	7 ②	4 ①
$R_2 = 0$	5 ①	2 X	5 X	3 X
$R_3 = -1$	2 (–1)	4 ③	5 ①	2 X

Entering cell (3, 1), exiting cell (2, 2)

From \ To	P	Q	R	S	Supply
A	4 40	2 40	7 –	4 –	80
B	5 –	2 –	5 40	3 70	110
C	2 30	4 –	5 –	2 30	60
Demand	70	40	40	100	250

	$K_1 = 4$	$K_2 = 2$	$K_3 = 6$	$K_4 = 4$
$R_1 = 0$	4 X	2 X	7 ①	4 ⓪
$R_2 = -1$	5 ②	2 ①	5 X	3 X
$R_3 = -2$	2 X	4 ④	5 ①	2 X

Allocation is optimal
Total cost = £770

From \ To	P	Q	R	S	Supply
A	4 10	2 40	7 –	4 30	80
B	5 –	2 –	5 40	3 70	110
C	2 60	4 –	5 –	2 –	60
Demand	70	40	40	100	250

Alternative allocation

Exercise 2.4

1 a, b

From \ To	X	Y	Dummy	Supply
A	4 120	7 80	0 –	200
B	3 –	5 70	0 30	100
Demand	120	150	30	300

	$K_1 = 4$	$K_2 = 7$	$K_3 = 2$
$R_1 = 0$	4 X	7 X	0 (–2)
$R_2 = -2$	3 ①	5 X	0 X

Revised allocation shown is optimal.
Cost = $120 \times 4 + 50 \times 7 + 100 \times 5 = £1330$

c

From \ To	X	Y	Dummy	Supply
A	4 120	7 50	0 30	200
B	3 –	5 100	0 –	100
Demand	120	150	30	300

Supplier A has 30 units left. Supplier B has no units left.

2 a, b

From \ To	X	Y	Z	Supply
A	4 260	3 –	4 –	260
B	5 40	2 280	6 –	320
C	7 –	3 80	5 110	190
Dummy	0 –	0 –	0 130	130
Demand	300	360	240	900

	$K_1 = 4$	$K_2 = 1$	$K_3 = 3$
$R_1 = 0$	4 X	3 ②	4 ①
$R_2 = 1$	5 X	2 X	6 ②
$R_3 = 2$	7 ①	3 X	5 X
$R_4 = -3$	0 ⑴	0 ②	0 X

Entering cell $(4, 1)$, exiting cell $(2, 1)$

From \ To	X	Y	Z	Supply
A	4 260	3 –	4 –	260
B	5 –	2 320	6 –	320
C	7 –	3 40	5 150	190
Dummy	0 40	0 –	0 90	130
Demand	300	360	240	900

	$K_1 = 4$	$K_2 = 2$	$K_3 = 4$
$R_1 = 0$	4 X	3 ①	4 ⓪
$R_2 = 0$	5 ①	2 X	6 ②
$R_3 = 1$	7 ②	3 X	5 X
$R_4 = -4$	0 X	0 ②	0 X

This allocation is optimal.
Alternative optimal allocation is

From \ To	X	Y	Z	Supply
A	4 170	3 –	4 90	260
B	5 –	2 320	6 –	320
C	7 –	3 40	5 150	190
Dummy	0 130	0 –	0 –	130
Demand	300	360	240	900

The minimum cost is £2550.

c **i** X must have shortfall of at least 40 units.
   ii Y will receive all the units needed.

3 **a**

From \ To	X	Y	Dummy	Supply
A	50 80	70 10	0 –	90
B	30 –	10 50	0 20	70
C	60 –	40 –	0 40	40
Demand	80	60	60	200

	$K_1 = 5$	$K_2 = 9$	$K_3 = 8$
$R_1 = 0$	5 X	9 X	6 ⑵
$R_2 = -3$	3 ①	6 X	5 X
$R_3 = -5$	4 ④	7 ③	3 X

Entering cell $(1, 3)$, exiting cell $(1, 2)$

From \ To	X	Y	Dummy	Supply
A	50 80	70 –	0 10	90
B	30 –	10 60	0 10	70
C	60 –	40 –	0 40	40
Demand	80	60	60	200

	$K_1 = 5$	$K_2 = 7$	$K_3 = 6$
$R_1 = 0$	5 X	9 ②	6 X
$R_2 = -1$	3 ⑴	6 X	5 X
$R_3 = -3$	4 ②	7 ③	3 X

Entering cell $(2, 1)$, exiting cell $(2, 3)$

From \ To	X	Y	Dummy	Supply
A	50 70	70 –	0 20	90
B	30 10	10 60	0 –	70
C	60 –	40 –	0 40	40
Demand	80	60	60	200

	$K_1 = 5$	$K_2 = 8$	$K_3 = 6$
$R_1 = 0$	5 — X	9 — (1)	6 — X
$R_2 = -2$	3 — X	6 — X	5 — (1)
$R_3 = -3$	4 — (2)	7 — (2)	3 — X

Allocation is optimal. Cost = £4400.

b A has 20 pallets left, C has 40.

4 a

From \ To	X	Y	Supply
A	10 — 75	7 — 9	84
B	4 — –	8 — 56	56
Dummy	0 — –	0 — 29	29
Demand	75	94	169

b

	$K_1 = 10$	$K_2 = 7$
$R_1 = 0$	10 — X	7 — X
$R_2 = 1$	4 — (−7)	8 — X
$R_3 = -7$	0 — (−3)	0 — X

Negative indices, so solution is not optimal.

c Entering cell $(2,1)$, exiting cell $(2,2)$

From \ To	X	Y	Supply
A	10 — 19	7 — 65	84
B	4 — 56	8 — –	56
Dummy	0 — –	0 — 29	29
Demand	75	94	169

	$K_1 = 10$	$K_2 = 7$
$R_1 = 0$	10 — X	7 — X
$R_2 = -6$	4 — X	8 — (7)
$R_3 = -7$	0 — (−3)	0 — X

Not yet optimal
Entering cell $(3,1)$, exiting cell $(1,1)$

From \ To	X	Y	Supply
A	10 — –	7 — 84	84
B	4 — 56	8 — –	56
Dummy	0 — 19	0 — 10	29
Demand	75	94	169

	$K_1 = 7$	$K_2 = 7$
$R_1 = 0$	10 — (3)	7 — X
$R_2 = -3$	4 — X	8 — (4)
$R_3 = -7$	0 — X	0 — X

Allocation is optimal.

d X is 19 workers short, Y is 10 workers short.

5 a

From \ To	X	Y	Z	Supply
A	3 — 130	3 — 50	3 — –	180
B	4 — –	2 — 100	6 — –	100
C	6 — –	5 — 30	4 — 140	170
Dummy	0 — –	0 — –	0 — 60	60
Demand	130	180	200	510

	$K_1 = 3$	$K_2 = 3$	$K_3 = 2$
$R_1 = 0$	3 — X	3 — X	3 — (1)
$R_2 = -1$	4 — (2)	2 — X	6 — (5)
$R_3 = 2$	6 — (1)	5 — X	4 — X
$R_4 = -2$	0 — (−1)	0 — (−1)	0 — X

Entering cell $(4,2)$, exiting cell $(3,2)$

D2

b

From \ To	X	Y	Z	Supply
A	3 — 130	3 — 50	3 — –	180
B	4 — –	2 — 100	6 — –	100
C	6 — –	5 — –	4 — 170	170
Dummy	0 — –	0 — 30	0 — 30	60
Demand	130	180	200	510

	$K_1 = 3$	$K_2 = 3$	$K_3 = 3$
$R_1 = 0$	3 X	3 X	3 (0)
$R_2 = -1$	4 (2)	2 X	6 (4)
$R_3 = 1$	6 (2)	5 (1)	4 X
$R_4 = -3$	0 (0)	0 X	0 X

This is an optimal solution (there are others, because of the zero indices in the table).
Cost = £142

c

From \ To	X	Y	Z	Supply
A	3 — 110	3 — 60	3 — 10	180
B	4 — –	2 — 100	6 — –	100
C	6 — –	5 — –	4 — 170	170
Dummy	0 — 20	0 — 20	0 — 20	60
Demand	130	180	200	510

Exercise 2.5

1 a

From \ To	X	Y	Z	Supply
A	3 — 50	1 — –	4 — –	50
B	2 — 10	4 — 30	3 — –	40
C	1 — –	5 — –	2 — 80	80
Demand	60	30	80	170

b A 3×3 table needs $(3 + 3 - 1) = 5$ entries. This has only 4, so is degenerate.

c Create extra occupied cell $(1,3)$. (Could be any except $(1,2)$)

	$K_1 = 3$	$K_2 = 5$	$K_3 = 4$
$R_1 = 0$	3 X	1 (-4)	4 X
$R_2 = -1$	2 X	4 X	3 (0)
$R_3 = -2$	1 (0)	5 (2)	2 X

Entering cell $(1,2)$, exiting cell $(2,2)$
Revised allocation is

From \ To	X	Y	Z	Supply
A	3 — 20	1 — 30	4 — 0	50
B	2 — 40	4 — –	3 — –	40
C	1 — –	5 — –	2 — 80	80
Demand	60	30	80	170

	$K_1 = 3$	$K_2 = 1$	$K_3 = 4$
$R_1 = 0$	3 X	1 X	4 X
$R_2 = -1$	2 X	4 (4)	3 (0)
$R_3 = -2$	1 (0)	5 (6)	2 X

This is optimal. Total cost = £330

2 a

From \ To	X	Y	Z	Supply
A	9 — 110	8 — 40	11 — –	150
B	8 — –	6 — 160	10 — –	160
C	4 — –	4 — –	7 — 150	150
Demand	110	200	150	460

b A 3×3 table needs $(3 + 3 - 1) = 5$ entries. This has only 4, so is degenerate.

c An extra cell at $(2,1)$ would leave $(3,3)$ as the only occupied cell in both row 3 and column 3, so impossible to find the shadow costs.

d

	$K_1 = 9$	$K_2 = 8$	$K_3 = 12$
$R_1 = 0$	9 X	8 X	11 (-1)
$R_2 = -2$	8 (1)	6 X	10 (0)
$R_3 = -5$	4 X	4 (1)	7 X

Entering cell $(1,3)$, exiting cell $(1,1)$

D2

From \ To	X	Y	Z	Supply
A	9 –	8 40	11 110	150
B	8 –	6 160	10 –	160
C	4 110	4 –	7 40	150
Demand	110	200	150	460

	$K_1 = 8$	$K_2 = 8$	$K_3 = 11$
$R_1 = 0$	9 (1)	8 X	11 X
$R_2 = -2$	8 (2)	6 X	10 (1)
$R_3 = -4$	4 X	4 (0)	7 X

This is optimal.

e Total cost = 3210

3 a

From \ To	X	Y	Z	Supply
A	7 50	5 30	8 –	80
B	9 –	6 100	9 –	100
C	3 –	2 50	4 90	140
D	5 –	2 –	3 70	70
Demand	50	180	160	390

b

From \ To	X	Y	Z	Supply
A	7 –	5 80	8 –	80
B	9 –	6 100	9 –	100
C	3 50	2 –	4 90	140
D	5 –	2 –	3 70	70
Demand	50	180	160	390

Entering cell $(3,1)$, exiting cells $(1,1)$ and $(3,2)$

c Should have $(4+3-1) = 6$ occupied cells. Only five remain, so degenerate.

d

	$K_1 = 7$	$K_2 = 5$	$K_3 = 8$
$R_1 = 0$	7 X	5 X	8 (0)
$R_2 = 1$	9 (1)	6 X	9 (0)
$R_3 = -4$	3 X	2 (1)	4 X
$R_4 = -5$	5 (3)	2 (2)	3 X

Create extra occupied cell e.g. $(1,1)$.
No negative indices, so optimal.

4 a

From \ To	X	Y	Z	Supply
A	6 70	4 20	7 –	90
B	8 –	5 70	6 25	95
Dummy	0 –	0 –	0 80	80
Demand	70	90	105	265

b

From \ To	X	Y	Z	Supply
A	6 –	4 90	7 –	90
B	8 –	5 –	6 95	95
Dummy	0 70	0 –	0 10	80
Demand	70	90	105	265

Entering cell $(3,1)$, exiting cells $(1,1)$ and $(2,2)$

c Should have $(3+3-1) = 5$ occupied cells.
Only four remain, so degenerate.

d

	$K_1 = 6$	$K_2 = 4$	$K_3 = 6$
$R_1 = 0$	6 X	4 X	7 (1)
$R_2 = 0$	8 (2)	5 (1)	6 X
$R_3 = -6$	0 X	0 (2)	0 X

Extra occupied cell, e.g. $(1,1)$
No negative indices, so optimal.

e X receives nothing. Z is 10 units short.
Y receives complete order.

D2

5 a

To / From	P	Q	R	S	Supply
A	[2] 60	[5] 20	[3] –	[4] –	80
B	[2] –	[4] 50	[1] –	[4] –	50
C	[5] –	[6] –	[2] 100	[3] –	100
D	[3] –	[7] –	[5] –	[5] 30	30
Demand	60	70	100	30	260

Should have $(4 + 4 - 1) = 7$ occupied cells. Only 5 are occupied, so degenerate.

b Extra occupied cells e.g. $(2,3)$ and $(3,4)$.
Entering cell $(4,1)$, exiting cell $(4,4)$

To / From	P	Q	R	S	Supply
A	[2] 30	[5] 50	[3] –	[4] –	80
B	[2] –	[4] 20	[1] 30	[4] –	50
C	[5] –	[6] –	[2] 70	[3] 30	100
D	[3] 30	[7] –	[5] –	[5] –	30
Demand	60	70	100	30	260

	$K_1 = 2$	$K_2 = 5$	$K_3 = 2$	$K_4 = 3$
$R_1 = 0$	[2] X	[5] X	[3] ①	[4] ①
$R_2 = -1$	[2] ①	[4] X	[1] X	[4] ②
$R_3 = 0$	[5] ③	[6] ①	[2] X	[3] X
$R_4 = 1$	[3] X	[7] ①	[5] ②	[5] ①

No negative indices, so optimal.

c The solution is unique.

Exercise 2.6

1 Minimise $C = 25x_{11} + 18x_{12} + 16x_{21} + 23x_{22}$
subject to $x_{11} + x_{12} = 540$
$x_{21} + x_{22} = 480$
$x_{11} + x_{21} = 390$
$x_{12} + x_{22} = 630$
$x_{11} \geqslant 0, x_{12} \geqslant 0, x_{21} \geqslant 0, x_{22} \geqslant 0$

2 a Minimise $4x_{11} + 2x_{12} + 5x_{13} + 3x_{21} + x_{22} + 4x_{23}$
$\qquad + 6x_{31} + 5x_{32} + 7x_{33}$
subject to $x_{11} + x_{12} + x_{13} = 250$
$x_{21} + x_{22} + x_{23} = 180$
$x_{31} + x_{32} + x_{33} = 140$
$x_{11} + x_{21} + x_{31} = 200$
$x_{12} + x_{22} + x_{32} = 220$
$x_{13} + x_{23} + x_{33} = 150$
$x_{11}, x_{12}, x_{13}, x_{21}, x_{22}, x_{23}, x_{31}, x_{32}, x_{33} \geqslant 0$

b The problem is balanced so values satisfying five equations automatically satisfy the sixth (equation [6] is [1] + [2] + [3] − [4] − [5])

3 Minimise $2x_{11} + 3x_{12} + 2x_{13} + 4x_{21} + 2x_{22} + 3x_{23}$
$\qquad + x_{31} + 5x_{32} + 4x_{33}$

subject to $x_{11} + x_{12} + x_{13} \leqslant 240$
$x_{21} + x_{22} + x_{23} \leqslant 200$
$x_{31} + x_{32} + x_{33} \leqslant 260$
$x_{11} + x_{21} + x_{31} = 180$
$x_{12} + x_{22} + x_{32} = 210$
$x_{13} + x_{23} + x_{33} = 220$
$x_{11}, x_{12}, x_{13}, x_{21}, x_{22}, x_{23}, x_{31}, x_{32}, x_{33} \geqslant 0$

Review 2

1 a

To / From	X	Y	Z	Supply
A	[3] 160	[5] –	[4] –	160
B	[4] 30	[3] 190	[2] 30	250
C	[5] –	[6] –	[5] 200	200
Demand	190	190	230	610

Total cost = 2230

b

	$K_1 = 3$	$K_2 = 2$	$K_3 = 1$
$R_1 = 0$	[3] X	[5] ③	[4] ③
$R_2 = 1$	[4] X	[3] X	[2] X
$R_3 = 4$	[5] –2	[6] ⓪	[5] X

There is a negative index, so solution is not optimal.

c Entering cell $(3,1)$, exiting cell $(2,1)$

To / From	X	Y	Z	Supply
A	[3] 160	[5] –	[4] –	160
B	[4] –	[3] 190	[2] 60	250
C	[5] 30	[6] –	[5] 170	200
Demand	190	190	230	610

	$K_1 = 3$	$K_2 = 4$	$K_3 = 3$
$R_1 = 0$	[3] X	[5] ①	[4] ①
$R_2 = -1$	[4] ②	[3] X	[2] X
$R_3 = 2$	[5] X	[6] ⓪	[5] X

No negative indices, so optimal. Total cost = 2170

2 a Total supply exceeds total demand. Problem is unbalanced, so dummy column needed.

b

From \ To	X	Y	Z	Dummy	Supply
A	15 / 8	20 / 4	24 / –	0 / –	12
B	12 / –	16 / 2	10 / 7	0 / –	9
C	28 / –	25 / –	25 / 3	0 / 5	8
Demand	8	6	10	5	29

c

	$K_1 = 15$	$K_2 = 20$	$K_3 = 14$	$K_4 = -11$
$R_1 = 0$	15 / X	20 / X	24 / (10)	0 / (11)
$R_2 = -4$	12 / (1)	16 / X	10 / X	0 / (15)
$R_3 = 11$	28 / (2)	25 / (–6)	25 / X	0 / X

Entering cell $(3,2)$, exiting cell $(2,2)$.

From \ To	X	Y	Z	Dummy	Supply
A	15 / 8	20 / 4	24 / –	0 / –	12
B	12 / –	16 / –	10 / 9	0 / –	9
C	28 / –	25 / 2	25 / 1	0 / 5	8
Demand	8	6	10	5	29

d

	$K_1 = 15$	$K_2 = 20$	$K_3 = 20$	$K_4 = -5$
$R_1 = 0$	15 / X	20 / X	24 / (4)	0 / (5)
$R_2 = -10$	12 / (7)	16 / (6)	10 / X	0 / (15)
$R_3 = 5$	28 / (8)	25 / X	25 / X	0 / X

No negative indices, so optimal.
Total cost = £365

e There are 5 unused cars left at garage C.

3 a, b

From \ To	Standard	Senior citizen	Dummy	Supply
Thursday	10 / 30	6 / –	0 / –	30
Friday	15 / 10	12 / 18	0 / –	28
Saturday	18 / –	12 / 7	0 / 15	22
Demand	40	25	15	80

c

	$K_1 = 10$	$K_2 = 7$	$K_3 = -5$
$R_1 = 0$	10 / X	6 / (–1)	0 / (5)
$R_2 = 5$	15 / X	12 / X	0 / (0)
$R_3 = 5$	18 / (3)	12 / X	0 / X

Negative index, so not optimal

d Entering cell $(1,2)$, exiting cell $(2,2)$

From \ To	Standard	Senior citizen	Dummy	Supply
Thursday	10 / 12	6 / 18	0 / –	30
Friday	15 / 28	12 / –	0 / –	28
Saturday	18 / –	12 / 7	0 / 15	22
Demand	40	25	15	80

	$K_1 = 10$	$K_2 = 6$	$K_3 = -6$
$R_1 = 0$	10 / X	6 / X	0 / (6)
$R_2 = 5$	15 / X	12 / (1)	0 / (1)
$R_3 = 6$	18 / (2)	12 / X	0 / X

No negative indices, so optimal.

4 a

From \ To	X	Y	Z	Supply
A	5 / 45	3 / –	2 / –	45
B	7 / 15	6 / 27	3 / –	42
C	9 / –	6 / –	2 / 18	18
Demand	60	27	18	105

b Should be $(3 + 3 - 1) = 5$ occupied cells. There are only 4, so degenerate.

c Extra occupied cell e.g. $(1,3)$

	$K_1 = 5$	$K_2 = 4$	$K_3 = 2$
$R_1 = 0$	5 / X	3 / (–1)	2 / X
$R_2 = 2$	7 / X	6 / X	3 / (–1)
$R_3 = 0$	9 / (4)	6 / (2)	2 / X

Negative index, so not optimal

D2

d Entering cell(1,2), exiting cell(2,2)

From \ To	X	Y	Z	Supply
A	[5] 18	[3] 27	[2] 0	45
B	[7] 42	[6] –	[3] –	42
C	[9] –	[6] –	[2] 18	18
Demand	60	27	18	105

	$K_1 = 5$	$K_2 = 3$	$K_3 = 2$
$R_1 = 0$	[5] X	[3] X	[2] (X)
$R_2 = 2$	[7] X	[6] (1)	[3] (−1)
$R_3 = 0$	[9] (4)	[6] (3)	[2] X

Still not optimal. Entering cell (2,3), exiting cell (1,3)

From \ To	X	Y	Z	Supply
A	[5] 18	[3] 27	[2] –	45
B	[7] 42	[6] –	[3] 0	42
C	[9] –	[6] –	[2] 18	18
Demand	60	27	18	105

	$K_1 = 5$	$K_2 = 3$	$K_3 = 1$
$R_1 = 0$	[5] X	[3] X	[2] (1)
$R_2 = 2$	[7] X	[6] (1)	[3] X
$R_3 = 1$	[9] (3)	[6] (2)	[2] X

No negative indices, so optimal.

5 Minimise $C = 3x_{11} + 5x_{12} + 4x_{13} + 4x_{21} + 3x_{22} + 2x_{23} + 5x_{31} + 6x_{32} + 5x_{33}$

subject to
$x_{11} + x_{12} + x_{13} = 160$
$x_{21} + x_{22} + x_{23} = 250$
$x_{31} + x_{32} + x_{33} = 200$
$x_{11} + x_{21} + x_{31} = 190$
$x_{12} + x_{22} + x_{32} = 190$
$x_{13} + x_{23} + x_{33} = 230$
$x_{11}, x_{12}, x_{13}, x_{21}, x_{22}, x_{23}, x_{31}, x_{32}, x_{33} \geqslant 0$

Chapter 3

Exercise 3.1

1 a

	X	Y	Z
A	4	0	5
B	0	0	6
C	5	1	0

Assignment A-Y, B-X, C-Z

b

	P	Q	R	S
A	6	2	0	5
B	0	0	3	1
C	1	9	4	0
D	0	5	5	1

Assignment A-R, B-Q, C-S, D-P

c

	X	Y	Z
A	0	0	4
B	2	0	3
C	5	4	0

Assignment A-X, B-Y, C-Z

d

	P	Q	R	S
A	4	0	2	3
B	7	5	0	0
C	2	0	1	6
D	0	3	4	7

No assignment possible.
R and S only assignable to B

2 a, c

	P	Q	R	S
A	0	0	4	1
B	3	2	0	6
C	5	0	9	0
D	1	2	0	0

Assignment A-P, B-R, C-Q, D-S

b, c

	P	Q	R	S
A	0	4	5	5
B	2	5	0	9
C	1	0	6	0
D	0	5	0	3

Assignment not possible.
A-P, B-R means no task for D

Exercise 3.2

1 a

	P	Q	R	S
A	4	0	3	1
B	2	0	5	0
C	1	3	2	0
D	6	0	0	3

No assignment possible

b

	P	Q	R	S
A	8	5	0	0
B	0	3	4	7
C	3	0	0	2
D	9	5	0	7

Assignment A-S, B-P, C-Q, D-R

c

	P	Q	R	S	T
A	0	7	0	2	0
B	4	0	8	8	2
C	0	3	3	0	4
D	5	0	2	0	9
E	1	0	3	5	2

No assignment possible

d

	P	Q	R	S	T
A	0	2	5	4	7
B	1	1	0	3	4
C	1	0	2	0	8
D	3	0	3	2	0
E	9	6	6	0	0

Assignment A-P, B-R, C-Q, D-T, E-S or A-P, B-R, C-S, D-Q, E-T

2 a

	Front crawl	Breast stroke	Back crawl	Butterfly
George	0	9	4	3
Harry	0	0	0	0
Ian	0	10	5	3
John	0	11	2	3

b

	Front crawl	Breast stroke	Back crawl	Butterfly
George	0	6	2	0
Harry	3	0	1	0
Ian	0	7	3	0
John	0	8	0	0

c George-Front crawl, Harry-Breast stroke, Ian-Butterfly, John-Back crawl or George-Butterfly, Harry-Breast stroke, Ian-Front crawl, John-Back crawl

d 193 s

3 a

	1	2	3
A	85	82	88
B	79	74	82
C	83	87	83

b

	1	2	3
A	3	0	6
B	5	0	8
C	0	4	0

c Revised matrix is

	1	2	3
A	0	0	3
B	2	0	5
C	0	7	0

Assign A-1, B-2, C-3. Total cost £24 200

4 a

	P	Q	R	S
A	15	25	32	40
B	27	18	20	36
C	12	24	29	13
D	15	32	39	23

b

	P	Q	R	S
A	0	10	15	24
B	9	0	0	17
C	0	12	15	0
D	0	17	22	7

Three lines used so no assignment possible

c Revised matrix is

	P	Q	R	S
A	0	0	5	24
B	19	0	0	27
C	0	2	5	0
D	0	7	12	7

(Other revisions possible)
Assignment A-Q, B-R, C-S, D-P

5 Opportunity cost matrix is

	P	Q	R	S	T
A	3	6	0	6	3
B	0	5	4	0	4
C	1	0	0	3	0
D	0	6	1	2	5
E	2	2	0	3	5

Revised matrix using lines shown is

	P	Q	R	S	T
A	3	4	0	4	1
B	2	5	6	0	4
C	3	0	2	3	0
D	0	4	1	0	3
E	2	0	0	1	3

Assignment A-R, B-S, C-T, D-P, E-Q.
Total cost = 54 days.

D2

179

6 a Opportunity cost matrix is

	1	2	3	4
A	5	60	20	0
B	0	0	0	0
C	0	30	*	0
D	0	35	15	0

b Revised matrix using the lines shown is

	1	2	3	4
A	5	45	5	0
B	15	0	0	15
C	0	15	*	0
D	0	20	0	0

Must assign A-4 and B-2. This forces D-3 and then C-1.

c £565 000

Exercise 3.3

1 a

	A	B	C	D	E
Ms Pugh	32	18	29	20	23
Mr Quentin	36	22	24	22	38
Dr Rani	20	18	21	12	29
Mrs Stuart	27	19	12	8	24
Miss Till	16	12	11	0	22

b

	A	B	C	D	E
Ms Pugh	6	0	9	2	0
Mr Quentin	6	0	0	0	11
Dr Rani	0	6	7	0	12
Mrs Stuart	11	11	2	0	11
Miss Till	8	12	9	0	17

c Revised matrix using the lines shown is

	A	B	C	D	E
Ms Pugh	8	0	9	4	0
Mr Quentin	8	0	0	2	11
Dr Rani	0	4	5	0	10
Mrs Stuart	11	9	0	0	9
Miss Till	8	10	7	0	15

Assignment Pugh-E, Quentin-B, Rani-A, Stuart-C, Till-D

2 a Cost matrix is

	Shot	Javelin	Discus	Hammer
Lisa	42	5	14	17
Marilyn	44	4	13	20
Nicola	40	8	16	18
Olive	41	0	9	24

Opportunity cost matrix is

	Shot	Javelin	Discus	Hammer
Lisa	5	0	1	2
Marilyn	8	0	1	6
Nicola	0	0	0	0
Olive	9	0	1	14

b Revised matrix is

	Shot	Javelin	Discus	Hammer
Lisa	3	0	0	0
Marilyn	6	0	0	4
Nicola	0	2	1	0
Olive	7	0	0	12

Assignments: Lisa-Hammer, Marilyn-Javelin, Nicola-Shot, Olive-Discus or Lisa-Hammer, Marilyn-Discus, Nicola-Shot, Olive-Javelin

c Expected score 146

Exercise 3.4

1 Introduce dummy column:

Language / Teacher	French	German	Spanish	Dummy
Mrs Atkins	7.2	6.4	10.4	0
Mr Beaumont	8.5	9.6	9.2	0
Ms Caldwell	12.2	7.5	7.8	0
Mr Delgado	8.8	9.3	7	0

Opportunity cost matrix is

Language / Teacher	French	German	Spanish	Dummy
Mrs Atkins	0	0	3.4	0
Mr Beaumont	1.3	3.2	2.2	0
Ms Caldwell	5	1.1	0.8	0
Mr Delgado	1.6	2.9	0	0

Revised matrix using lines shown

Language / Teacher	French	German	Spanish	Dummy
Mrs Atkins	0	0	4.5	1.1
Mr Beaumont	0.2	2.1	2.2	0
Ms Caldwell	3.9	0	0.8	0
Mr Delgado	0.5	1.8	0	0

Assign Atkins-French, Caldwell-German, Delgado-Spanish. Beaumont not assigned.

2 **a** Opportunity cost matrix:

Worker \ Location	1	2	3	4	5
A	1	5	3	0	4
B	0	3	7	8	2
C	0	1	3	5	3
D	3	5	0	2	8
Dummy	0	0	0	0	0

Zeros can be covered with 4 lines, so assignment not possible.

b Revised matrix is

Worker \ Location	1	2	3	4	5
A	1	4	3	0	3
B	0	2	7	8	1
C	0	0	3	5	2
D	3	4	0	2	7
Dummy	1	0	1	1	0

Allocate A-4, B-1, C-2, D-3. No worker for location 5.

c Travel cost for week = £92

3 **a**

Winner \ Production	1	2	3	4	5
A	7	2	5	4	5
B	2	6	5	3	6
C	1	3	3	0	4
D	4	2	6	2	4
Dummy	0	0	0	0	0

b The opportunity cost matrix is

Winner \ Production	1	2	3	4	5
A	5	0	3	2	3
B	0	4	3	1	4
C	1	3	3	0	4
D	2	0	4	0	2
Dummy	0	0	0	0	0

Allocation not yet possible.

Revised matrix is

Winner \ Production	1	2	3	4	5
A	5	0	1	2	1
B	0	4	1	1	2
C	1	3	1	0	2
D	2	0	2	0	0
Dummy	2	2	0	2	0

Allocate A-2, B-1, C-4, D-5

c No-one attends production 3.

Exercise 3.5

1 **a** The decision variables are x_{ij}, each taking the values 0 or 1 depending on whether cell (i, j) is in or not in the allocation.

b $x_{11} + x_{12} + x_{13} = 1$
$x_{21} + x_{22} + x_{23} = 1$
$x_{31} + x_{32} + x_{33} = 1$
$x_{11} + x_{21} + x_{31} = 1$
$x_{12} + x_{22} + x_{32} = 1$
$x_{13} + x_{23} + x_{33} = 1$

c Minimise $C = 5x_{11} + 12x_{12} + 7x_{13} + 6x_{21} + 11x_{22} + 6x_{23} + 10x_{31} + 14x_{32} + 9x_{33}$

2 Maximise $T = 86x_{11} + 72x_{12} + 67x_{13} + 58x_{21} + 66x_{22} + 60x_{23} + 74x_{31} + 76x_{32} + 80x_{33}$

subject to $x_{11} + x_{12} + x_{13} = 1$
$x_{21} + x_{22} + x_{23} = 1$
$x_{31} + x_{32} + x_{33} = 1$
$x_{11} + x_{21} + x_{31} = 1$
$x_{12} + x_{22} + x_{32} = 1$
$x_{13} + x_{23} + x_{33} = 1$
Each $x_{ij} = 0$ or 1

3 **a** 25 **b** 10 constraints, 9 of them independent.

Review 3

1 **a**

	Bob	Carol	Deepak
Luxicabs	0	0	0
Maxitaxi	1.5	0	1
Nobby's	3	0	0.5

b Zeros can be covered with two lines, as shown. Should be three lines for an allocation to be possible.

c Revised matrix is

	Bob	Carol	Deepak
Luxicabs	0	0.5	0
Maxitaxi	1	0	0.5
Nobby's	2.5	0	0

d Bob in Luxicabs, Carol in Maxitaxi and Deepak in Nobby's. Total cost £22.

D2

2 a The problem is unbalanced. There must be the same number of columns and rows for an allocation to be made.

b Cost matrix is

	1	2	3	4	Dummy
A	2	9	3	7	0
B	7	13	0	10	0
C	5	9	0	8	0
D	9	8	1	7	0
E	4	7	5	5	0

Opportunity cost matrix is

	1	2	3	4	Dummy
A	0	2	3	2	0
B	5	6	0	5	0
C	3	2	0	3	0
D	7	1	1	2	0
E	2	0	5	0	0

c Revised matrix using the lines shown is

	1	2	3	4	Dummy
A	0	2	4	2	1
B	4	5	0	4	0
C	2	1	0	2	0
D	6	0	1	1	0
E	2	0	6	0	1

d Possible allocations are A-1, B-3, D-2, E-4 or A-1, C-3, D-2, E-4, so either B or C fails to get a house.

e £850 000

3 Minimise $C = 7.5x_{11} + 5x_{12} + 10x_{13} + 8.5x_{21} + 4.5x_{22} + 10.5x_{23} + 10x_{31} + 4.5x_{32} + 10x_{33}$

subject to $x_{11} + x_{12} + x_{13} = 1$
$x_{21} + x_{22} + x_{23} = 1$
$x_{31} + x_{32} + x_{33} = 1$
$x_{11} + x_{21} + x_{31} = 1$
$x_{12} + x_{22} + x_{32} = 1$
$x_{13} + x_{23} + x_{33} = 1$
Each $x_{ij} = 0$ or 1

4 a i Opportunity cost matrix is

Council \ Scheme	A	B	C	D
1	0.4	0	0	1.1
2	0	0	0.4	0
3	0.2	0	1.9	2.6
4	0.8	0	0.4	1.5

Revised matrix using the lines shown is

Council \ Scheme	A	B	C	D
1	0.4	0.2	0	1.1
2	0	0.2	0.4	0
3	0	0	1.7	2.4
4	0.6	0	0.2	1.3

Allocate 1-C, 2-D, 3-A, 4-B.

ii 284 jobs created at cost of £31.2 million = £109 859 per job.

b i Cost matrix is

Council \ Scheme	A	B	C	D
1	8	20	13	3
2	11	10	0	5
3	7	21	12	2
4	13	23	12	6

Opportunity cost matrix is

Council \ Scheme	A	B	C	D
1	0	7	10	0
2	6	0	0	5
3	0	9	10	0
4	2	7	6	0

Revised matrix using the lines shown is

Council \ Scheme	A	B	C	D
1	0	1	4	0
2	12	0	0	11
3	0	3	4	0
4	2	1	0	0

Allocate 1-A, 2-B, 3-D, 4-C or 1-D, 2-B, 3-A, 4-C

ii Both create 300 jobs. The second allocation is cheaper, at £32.7 million, which is £109 000 per job.

5   a   Cost matrix is

Plot / Vegetable	1	2	3	4	5
A	4	8	12	2	6
B	0	5	12	4	8
C	11	10	8	3	8
D	2	9	3	5	1
Dummy	0	0	0	0	0

Opportunity cost matrix is

Plot / Vegetable	1	2	3	4	5
A	2	6	10	0	4
B	0	5	12	4	8
C	8	7	5	0	5
D	1	8	2	4	0
Dummy	0	0	0	0	0

Revised matrix using the lines shown is

Plot / Vegetable	1	2	3	4	5
A	2	4	8	0	4
B	0	3	10	4	8
C	8	5	3	0	5
D	1	6	0	4	0
Dummy	2	0	0	2	2

Still only four lines needed, as shown.
Revised matrix is

Plot / Vegetable	1	2	3	4	5
A	2	1	5	0	1
B	0	0	7	4	5
C	8	2	0	0	2
D	1	6	0	7	0
Dummy	2	0	0	5	2

Allocate A-4, B-1, C-3, D-5.

b   Shed on plot 2.

Chapter 4

Exercise 4.1

1   a

		B_1	B_2	B_3	Row minima	Max of row minima
A	A_1	5	3	9	3	
	A_2	4	7	6	4	4
	A_3	2	4	5	2	
	Column maxima	5	7	9		
	Min of column maxima	5				

Play safe strategies A_2 and B_1. Not a stable solution.

b

		B_1	B_2	B_3	Row minima	Max of row minima
A	A_1	9	-2	3	-2	
	A_2	-1	-4	2	-4	
	A_3	4	6	3	3	3
	Column maxima	9	6	3		
	Min of column maxima			3		

Play safe strategies A_3 and B_3. Stable solution.
Value = 3.

c

		B_1	B_2	B_3	Row minima	Max of row minima
A	A_1	6	-1	4	-1	-1
	A_2	-1	-3	5	-3	
	Column maxima	6	-1	5		
	Min of column maxima		-1			

Play safe strategies A_1 and B_2. Stable solution.
Value = -1.

d

		B_1	B_2	Row minima	Max of row minima
A	A_1	-1	3	-1	
	A_2	2	7	2	2
	A_3	6	-2	-2	
	Column maxima	6	7		
	Min of column maxima	6			

Play safe strategies A_2 and B_1. Not a stable solution.

D2

e

		B					
		B_1	B_2	B_3	B_4	Row minima	Max of row minima
A	A_1	4	-2	5	9	-2	
	A_2	2	1	3	5	1	1
	A_3	3	-1	-8	4	-8	
	Column maxima	4	1	5	9		
	Min of column maxima		1				

Playsafe strategies A_2 and B_2. Stable solution. Value = 1.

f

		B					
		B_1	B_2	B_3	B_4	Row minima	Max of row minima
A	A_1	3	2	7	10	2	
	A_2	5	5	7	8	5	5
	A_3	-3	2	-5	6	-5	
	Column maxima	5	5	7	10		
	Min of column maxima	5	5				

Playsafe strategies A_2 and B_1 or B_2. Stable solution. Value = 5.

2

		A		
		A_1	A_2	A_3
B	B_1	1	-2	-6
	B_2	-3	-7	2

3 a

		B		
		B_1	B_2	B_3
A	A_1	2	0	-2
	A_2	-2	-6	0
	A_3	6	2	4

b

		B				
		B_1	B_2	B_3	Row minima	Max of row minima
A	A_1	1	0	-1	-1	
	A_2	-1	-3	0	-3	
	A_3	3	1	2	1	1
	Column maxima	3	1	2		
	Min of column maxima		1			

Playsafe strategies A_3 and B_2. Max of row minima = min of col maxima = 1, so stable solution (saddle point).

c Player A would win 10–0.

Exercise 4.2

1 a

		B	
		B_1	B_2
A	A_2	2	3
	A_3	4	-2

b

		B
		B_2
A	A_1	-1

This is stable. Play A_1 and B_2. Value = -1

c

		B	
		B_1	B_2
A	A_1	3	1
	A_2	2	5

d

		B	
		B_1	B_3
A	A_2	3	-2
	A_3	-4	1

e

		B	
		B_1	B_4
A	A_2	-1	-1

This is stable. Play A_2 and B_1 or B_4. Value = −1.

f

		B	
		B_1	B_3
A	A_1	3	2
	A_3	1	4

2 a i

		Y		
		Choose 2	Choose 6	Choose 7
X	Choose 3	-2	3	4
	Choose 5	3	-2	-4
	Choose 9	7	3	-4

ii

		Y		
		Choose 2	Choose 6	Choose 7
X	Choose 3	-2	3	4
	Choose 9	5	3	-4

Not stable because a stable solution would reduce to a single cell.

b e.g. Change Yolanda's set to {3, 6, 7}

Exercise 4.3

1 **a** Value $= 2\frac{2}{7}$. A plays A_1, A_2 with probabilities $\frac{1}{7}, \frac{6}{7}$.
B plays B_1, B_2 with probabilities $\frac{2}{7}, \frac{5}{7}$.

b Value $= 2\frac{1}{3}$. A plays $A_1 A_2$ with probabilities $\frac{2}{3}, \frac{1}{3}$.
B plays B_1, B_2 with probabilities $\frac{1}{6}, \frac{5}{6}$.

c This has a stable solution where A plays A_2 and B plays B_2. Value $= 2$.

d Value $= 3\frac{1}{4}$. A plays A_1, A_2 with probabilities $\frac{1}{4}, \frac{3}{4}$.
B plays B_1, B_2 with probabilities $\frac{1}{4}, \frac{3}{4}$.

Exercise 4.4

1 **a** A plays A_1, A_2 with probabilities $\frac{5}{9}, \frac{4}{9}$. B does not play B_1 and plays B_2, B_3 with probabilities $\frac{5}{9}, \frac{4}{9}$.
Value $= \frac{2}{9}$.

b A plays A_1, A_2 with probabilities $\frac{3}{4}, \frac{1}{4}$. B does not play B_3 and plays B_1, B_2 with probabilities $\frac{1}{2}, \frac{1}{2}$.
Value $= 6\frac{1}{2}$.

c A plays A_1, A_2 with probabilities $\frac{4}{7}, \frac{3}{7}$. B does not play B_3 and plays B_1, B_2 with probabilities $\frac{1}{7}, \frac{6}{7}$.
Value $= 2\frac{3}{7}$.

d A plays A_1, A_2 with probabilities $\frac{2}{5}, \frac{3}{5}$. B does not play B_2 and plays B_1, B_3 with probabilities $\frac{1}{5}, \frac{4}{5}$.
Value $= \frac{2}{5}$.

2 **a** B plays B_1, B_2 with probabilities $\frac{1}{2}, \frac{1}{2}$. A does not play A_1 and plays A_2, A_3 with probabilities $\frac{1}{8}, \frac{7}{8}$.
Value $= \frac{1}{2}$.

b B plays B_1, B_2 with probabilities $\frac{1}{3}, \frac{2}{3}$. A does not play A_1 and plays A_2, A_3 with probabilities $\frac{1}{3}, \frac{2}{3}$.
Value $= -1$.

c B plays B_1, B_2 with probabilities $\frac{1}{4}, \frac{3}{4}$. A does not play A_2 and plays A_1, A_3 with probabilities $\frac{1}{2}, \frac{1}{2}$. Value $= 7\frac{1}{2}$.

d B plays B_1, B_2 with probabilities $\frac{1}{2}, \frac{1}{2}$. A plays A_1, A_2 with probabilities $\frac{1}{2}, \frac{1}{2}$. A does not play A_3. Value $= \frac{-3}{2}$.

3 **a**

		B				Row minima	Max of row minima
		B_1	B_2	B_3	B_4		
	A_1	-1	3	7	-1	-1	
A	A_2	1	5	4	2	1	
	A_3	4	2	3	3	2	2
Column maxima		4	5	7	3		
Min of column maxima					3		

Max of row minima $\neq$ min of column maxima, so no saddle point.

b

		B		
		B_1	B_2	B_4
A	A_2	1	5	2
	A_3	4	2	3

c A plays A_1, A_2 with probabilities $\frac{1}{4}, \frac{3}{4}$. Value $= 2\frac{3}{4}$.

d B does not play B_1 and plays B_2, B_4 with probabilities $\frac{1}{4}, \frac{3}{4}$.

4 **a**

		Y				Row minima	Max of row minima
		I	II	III	IV		
X	I	1	4	3	2	1	1
	II	-1	3	0	3	-1	
	III	4	2	5	1	1	1
	IV	-2	6	-3	1	-3	
Column maxima		4	6	5	3		
Min of column maxima					3		

Max of row minima $\neq$ min of column maxima, so no saddle point.

b

		Y	
		I	IV
X	I	1	2
	II	-1	3
	III	4	1

c Y plays I, IV with probabilities $\frac{1}{4}, \frac{3}{4}$. X does not play II and plays I, III with probabilities $\frac{1}{4}, \frac{3}{4}$.
Value $= 1\frac{3}{4}$.

5 **a**

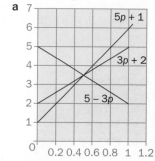

A's best strategy when $5p + 1 = 3p + 2 = 5 - 3p$, giving $p = \frac{1}{2}$.

b Value $= 3\frac{1}{2}$.

c All lines pass through the same point, so none of B's options is worse than the others.

d **i** $q_1 + q_2 + q_3 = 1$ [1]
$2q_1 + 5q_2 + 6q_3 = 3\frac{1}{2}$ [2]
$5q_1 + 2q_2 + q_3 = 3\frac{1}{2}$ [3]

ii From [3] − [1], $q_2 = 2\frac{1}{2} - 4q$. Sub in [1] gives $q_3 = 3q - 1\frac{1}{2}$

iii $\frac{1}{2} \leqslant q \leqslant \frac{5}{8}$ because $q_2, q_3 \geqslant 0$.
$0 \leqslant q_2 \leqslant \frac{1}{2}, 0 \leqslant q_3 \leqslant \frac{3}{8}$.

Exercise 4.5

1 a Maximise $P = v$
subject to $v - 5p_1 - 2p_2 - 3p_3 + r = 0$
$v - p_1 - 4p_2 - 2p_3 + s = 0$
$p_1 + p_2 + p_3 + t = 1$
or to $v - 2p_1 + p_2 + r = 3$
$v + p_1 - 2p_2 + s = 2$

b Add 5 to table entries. If V is the value of the original game, let $v = V + 5$
Maximise $P = v$
subject to $v - 2p_1 - 4p_2 - 5p_3 + r = 0$
$v - 6p_1 - p_2 + s = 0$
$p_1 + p_2 + p_3 + t = 1$
or to $v + 3p_1 + p_2 + r = 5$
$v - 6p_1 - p_2 + s = 0$

c Maximise $P = v$
subject to $v - 2p_1 - p_2 - 5p_3 + r = 0$
$v - p_1 - 4p_2 - 2p_3 + s = 0$
$v - 3p_1 - 4p_2 - p_3 + t = 0$
$p_1 + p_2 + p_3 + u = 1$
or to $v + 3p_1 + 4p_2 + r = 5$
$v + p_1 - 2p_2 + s = 2$
$v - 2p_1 - 3p_2 + t = 1$

d Add 8 to table entries. If V is the value of the original game, let $v = V + 8$
Maximise $P = v$
subject to $v - 5p_2 - 8p_3 + r = 0$
$v - 6p_1 - 7p_2 - 5p_3 + s = 0$
$v - 9p_1 - 4p_2 - 6p_3 + t = 0$
$p_1 + p_2 + p_3 + u = 1$
or to $v + 8p_1 + 3p_2 + r = 8$
$v - p_1 - 2p_2 + s = 5$
$v - 3p_1 + 2p_2 + t = 6$

2 a B plays strategies B_1, B_2 and B_3 with probabilities q_1, q_2 and q_3.
$$x_1 = \frac{q_1}{v}, x_2 = \frac{q_2}{v}, x_3 = \frac{q_3}{v},$$
where v is the value of the game.
Maximise $P = x_1 + x_2 + x_3$
subject to $5x_1 + 3x_2 + x_3 \leqslant 1$
$x_1 + 2x_2 + 6x_3 \leqslant 1$

b Add 5 to table entries. If V is the value of the original game, let $v = V + 5$
B plays strategies B_1, B_2 and B_3 with probabilities q_1, q_2 and q_3.
$$x_1 = \frac{q_1}{v}, x_2 = \frac{q_2}{v}, x_3 = \frac{q_3}{v}$$
Maximise $P = x_1 + x_2 + x_3$
subject to $x_1 + 3x_2 + 4x_3 \leqslant 1$
$4x_1 + 8x_2 \leqslant 1$

c B plays strategies B_1, B_2 and B_3 with probabilities q_1, q_2 and q_3.
$$x_1 = \frac{q_1}{v}, x_2 = \frac{q_2}{v}, x_3 = \frac{q_3}{v},$$
where v is the value of the game.
Maximise $P = x_1 + x_2 + x_3$
subject to $3x_1 + 3x_2 + 2x_3 \leqslant 1$
$x_1 + 5x_2 + 4x_3 \leqslant 1$
$4x_1 + 2x_2 + 3x_3 \leqslant 1$

d Add 5 to table entries. If V is the value of the original game, let $v = V + 5$
B plays strategies B_1, B_2 and B_3 with probabilities q_1, q_2 and q_3.
$$x_1 = \frac{q_1}{v}, x_2 = \frac{q_2}{v}, x_3 = \frac{q_3}{v}$$
Maximise $P = x_1 + x_2 + x_3$
subject to $3x_1 + 4x_2 + 6x_3 \leqslant 1$
$4x_1 + 2x_2 + x_3 \leqslant 1$
$5x_1 + 6x_3 \leqslant 1$

3 a i Add 2 to table entries. If V is the value of the original game, let $v = V + 2$
Maximise $P = v$
subject to $v + p_1 - p_2 \leqslant 2$
$v - p_1 + p_2 \leqslant 1$
$v + 3p_1 + 2p_2 \leqslant 3$

ii

Basic variable	v	p_1	p_2	s	t	u	Value	Row no.
s	1	1	-1	1	0	0	2	[1]
t	1	-1	1	0	1	0	1	[2]
u	1	3	2	0	0	1	3	[3]
P	-1	0	0	0	0	0	0	[4]

Basic variable	v	p_1	p_2	s	t	u	Value	Row no.
s	0	2	-2	1	-1	0	1	[5] = [1] − [6]
v	1	-1	1	0	1	0	1	[6] = [2]
u	0	4	1	0	-1	1	2	[7] = [3] − [6]
P	0	-1	1	0	1	0	1	[8] = [4] + [6]

Basic variable	v	p_1	p_2	s	t	u	Value	Row no.
p_1	0	1	-1	0.5	-0.5	0	0.5	$[9] = \dfrac{[5]}{2}$
v	1	0	0	0.5	0.5	0	1.5	[10] = [6] + [9]
u	0	0	5	-2	1	1	0	[11] = [7] − 4 × [9]
P	0	0	0	0.5	0.5	0	1.5	[12] = [8] + [9]

$v = 1.5$ when $p_1 = 0.5$, $p_2 = 0$ and so $p_3 = 0.5$.
The value of the game $= v - 2 = -0.5$

b i Add 2 to table entries. If V is the value of the original game, let $v = V + 2$
B plays strategies B_1, B_2 and B_3 with probabilities q_1, q_2 and q_3.
$$x_1 = \frac{q_1}{v}, x_2 = \frac{q_2}{v}, x_3 = \frac{q_3}{v}$$
Maximise $P = x_1 + x_2 + x_3$
subject to $x_1 + 2x_2 \leqslant 1$
$3x_1 + x_3 \leqslant 1$
$2x_1 + x_2 + 3x_3 \leqslant 1$

D2

ii

Basic variable	x_1	x_2	x_3	s	t	u	Value	Row no.
s	1	2	0	1	0	0	1	[1]
t	3	0	1	0	1	0	1	[2]
u	2	1	3	0	0	1	1	[3]
P	-1	-1	-1	0	0	0	0	[4]

Basic variable	x_1	x_2	x_3	s	t	u	Value	Row no.
s	0	2	$-\frac{1}{3}$	1	$-\frac{1}{3}$	0	$\frac{2}{3}$	$[5] = [1] - [6]$
x_1	1	0	$\frac{1}{3}$	0	$\frac{1}{3}$	0	$\frac{1}{3}$	$[6] = \frac{[2]}{3}$
u	0	1	$2\frac{1}{3}$	0	$-\frac{2}{3}$	1	$\frac{1}{3}$	$[7] = [3] - [6] \times 2$
P	0	-1	$-\frac{2}{3}$	0	$\frac{1}{3}$	0	$\frac{1}{3}$	$[8] = [4] + [6]$

Basic variable	x_1	x_2	x_3	s	t	u	Value	Row no.
s	0	0	-5	1	1	-2	0	$[9] = [5] - 2 \times [11]$
x_1	1	0	$\frac{1}{3}$	0	$\frac{1}{3}$	0	$\frac{1}{3}$	$[10] = [6]$
x_2	0	1	$2\frac{1}{3}$	0	$-\frac{2}{3}$	1	$\frac{1}{3}$	$[11] = [7]$
P	0	0	$-1\frac{2}{3}$	0	$-\frac{1}{3}$	1	$\frac{2}{3}$	$[12] = [8] + [11]$

Basic variable	x_1	x_2	x_3	s	t	u	Value	Row no.
t	0	0	-5	1	1	-2	0	$[13] = [9]$
x_1	1	0	2	$-\frac{1}{3}$	0	$\frac{2}{3}$	$\frac{1}{3}$	$[14] = [10] - \frac{[13]}{3}$
x_2	0	1	-1	$\frac{2}{3}$	0	$-\frac{1}{3}$	$\frac{1}{3}$	$[15] = [11] + 2 \times \frac{[13]}{3}$
P	0	0	0	$\frac{1}{3}$	0	$\frac{1}{3}$	$\frac{2}{3}$	$[16] = [12] + \frac{[13]}{3}$

$x_1 = x_2 = \frac{1}{3}$, $x_3 = 0$. $P = \frac{1}{v} = \frac{2}{3} \Rightarrow v = 1.5$, as before.

B plays B_1, B_2, B_3 with probabilities $\frac{1}{2}, \frac{1}{2}, 0$

Review 4

1 a In a zero-sum game the total gain on each play is zero. If it is a two-person game, one person's winnings equal the other person's losses.

b i

					Row minima	Max of row minima
		B				
		I	II	III		
	I	1	0	-2	-2	
A	II	-1	4	0	-1	
	III	2	3	5	2	2
	Column maxima	2	4	5		
	Min of column maxima	2				

A plays III, B plays I.

ii Max of row minima = min of col maxima = 2, so stable solution (saddle point).
Neither player gains by deviating from the playsafe strategy.

iii Value = 2

c

		A		
		I	II	III
	I	-1	0	2
B	II	1	-4	0
	III	-2	-3	-5

2 a

		B		Row minima	Max of row minima
		B_1	B_2		
A	A_1	-1	3	-1	
	A_2	4	2	2	2
	Column maxima	4	3		
	Min of column maxima		3		

Max of row minima ≠ min of column maxima, so no stable solution (saddle point)

b Value = $2\frac{1}{3}$. A plays A_1, A_2 with probabilities $\frac{1}{3}, \frac{2}{3}$.

c B plays B_1, B_2 with probabilities $\frac{1}{6}, \frac{5}{6}$

3 a

		B	
		I	II
A	I	5	3
	II	4	7

b

		B		Row minima	Max of row minima
		I	II		
A	I	5	3	3	
	II	4	7	4	4
	Column maxima	5	7		
	Min of column maxima	5			

Max of row minima ≠ min of column maxima, so no stable solution (saddle point)

c Value = $4\frac{3}{5}$. A plays I, II with probabilities $\frac{3}{5}, \frac{2}{5}$.
B plays I, II with probabilities $\frac{4}{5}, \frac{1}{5}$

4 a

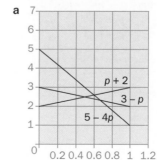

Value = $2\frac{1}{2}$. A plays A_1, A_2 with probabilities $\frac{1}{2}, \frac{1}{2}$

b The expected value $5 - 4p$ corresponds to B_3. The line passes above the maximum point of the feasible region, so using this would give A a better average gain. B plays B_1, B_2 with probabilities $\frac{1}{2}$, $\frac{1}{2}$.

5 a

			B			
			I	II	Row minima	Max of row minima
A	I		-2	4	-2	
	II		-1	2	-1	
	III		4	1	1	1
Column maxima			4	4		
Min of column maxima			4	4		

Max of row minima $\neq$ min of column maxima, so no stable solution (saddle point)

b

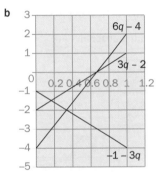

B plays I, II with probabilities $\frac{1}{6}$, $\frac{5}{6}$

c Strategy I corresponds to the expected value $6q - 4$, which passes below the minimum point of the feasible region, so B's returns would be improved. A plays II, III with probabilities $\frac{1}{6}$, $\frac{5}{6}$

d Value $= 1\frac{1}{2}$

6 a B plays B_1, B_2 with probabilities $\frac{1}{2}$, $\frac{1}{2}$

b Add 4 to table entries. If V is the value of the original game, let $v = V + 4$
Maximise $P = v$
subject to $v - 6p_1 - p_2 - 3p_3 + r = 0$
$v - 5p_2 - p_3 + s = 0$
$p_1 + p_2 + p_3 + t = 1$

or to $\quad v - 3p_1 + 2p_2 + r = 3$
$v + p_1 - 4p_2 + s = 1$

7 a

		N		
		I	II	III
M	I	-2	1	3
	II	2	-1	1
	III	-3	2	1

b Add 3 to table entries. If V is the value of the original game, let $v = V + 3$
B plays strategies B_1, B_2 and B_3 with probabilities q_1, q_2 and q_3.

$$x_1 = \frac{q_1}{v}, x_2 = \frac{q_2}{v}, x_3 = \frac{q_3}{v}$$

Maximise $\quad P = x_1 + x_2 + x_3$
subject to $\quad x_1 + 4x_2 + 6x_3 \leqslant 1$
$\qquad\qquad 5x_1 + 2x_2 + 4x_3 \leqslant 1$
$\qquad\qquad\quad 5x_2 + 4x_3 \leqslant 1$

Revision 1

1 a $x + 2y + 4z \leqslant 24$
b i $x + 2y + 4z + s = 24$
 ii s is the idle time on the machine
c 1 euro
d

Basic variable	After first iteration					
	x	y	z	r	s	Value
r	$\frac{3}{2}$	2	0	1	$-\frac{3}{2}$	14
z	$\frac{1}{4}$	$\frac{1}{2}$	1	0	$\frac{1}{4}$	6
P	0	-1	0	0	1	24
	After second iteration					
y	$\frac{3}{4}$	1	0	$\frac{1}{2}$	$-\frac{3}{4}$	7
z	$-\frac{1}{8}$	0	1	$-\frac{1}{4}$	$\frac{5}{8}$	$\frac{5}{2}$
P	$\frac{3}{4}$	0	0	$\frac{1}{2}$	$\frac{1}{4}$	31

$x = r = s = 0$, $y = 7$, $z = 2.5$. Profit 31 euros.
e Can't make half a lamp.
f e.g. $x = 0$, $y = 10$, $z = 0$ or $x = 0$, $y = 6$, $z = 3$ or $x = 1$, $y = 7$, $z = 2$

2 a Maximise $P = 4x + 5y + 3z$ subject to $3x + 2y + 4z \leqslant 35$, $x + 3y + 2z \leqslant 20$, $4x + 5y + 3z \leqslant 24$

b

Basic variable	x	y	z	r	s	t	Value
r	2	0	$\frac{5}{4}$	1	0	$-\frac{1}{2}$	23
s	$-\frac{1}{2}$	0	$-\frac{1}{4}$	0	1	$-\frac{3}{4}$	2
y	$\frac{1}{2}$	1	$\frac{3}{4}$	0	0	$\frac{1}{4}$	6
P	$-\frac{3}{2}$	0	$\frac{3}{4}$	0	0	$\frac{5}{4}$	30

Basic variable	x	y	z	r	s	t	Value
x	1	0	$\frac{5}{4}$	$\frac{1}{2}$	0	$-\frac{1}{4}$	$\frac{23}{2}$
s	0	0	$\frac{3}{8}$	$\frac{1}{4}$	1	$-\frac{7}{8}$	$\frac{31}{4}$
y	0	1	$\frac{1}{8}$	$-\frac{1}{4}$	0	$\frac{3}{8}$	$\frac{1}{4}$
P	0	0	$\frac{21}{8}$	$\frac{21}{8}$	0	$\frac{7}{8}$	$\frac{189}{4}$

$P = 47\frac{1}{4}$, $x = 11\frac{1}{2}$, $y = \frac{1}{4}$, $z = 0$

c There is slack on blending ($s = 7.75$) so do not alter this. Increase time available for processing and packing.

3 a No negative elements in the profit row.

b $P = 11, x = 1, y = \frac{1}{3}, z = 0; r = \frac{2}{3} s = 0, t = 0$

c $P + z + s + t = 11 \Rightarrow P = 11 - z - s - t$ so increasing z, s or t would decrease P.

4 a $P = 2x + 3y + 4z$

b $12x + 4y + 5z \leqslant 246, 9x + 6y + 3z \leqslant 153,$
$5x + 2y - 2z \leqslant 171$

c

Basic variable	x	y	z	r	s	t	Value	Row no.
r	12	4	5	1	0	0	246	[1]
s	9	6	3	0	1	0	153	[2]
t	5	2	-2	0	0	1	171	[3]
P	-2	-3	-4	0	0	0	0	[4]

Basic variable	x	y	z	r	s	t	Value	Row no.
r	6	0	3	1	$-\frac{2}{3}$	0	144	$[5] = [1] - [6] \times 4$
y	$\frac{3}{2}$	1	$\frac{1}{2}$	0	$\frac{1}{6}$	0	25.5	$[6] = \frac{[2]}{6}$
t	2	0	-3	0	$-\frac{1}{3}$	1	120	$[7] = [3] - [6] \times 2$
P	4	0	-1	0	$\frac{2}{3}$	0	102	$[8] = [4] + [6] \times 4$

Basic variable	x	y	z	r	s	t	Value	Row no.
z	2	0	1	$\frac{1}{3}$	$-\frac{2}{9}$	0	48	$[9] = \frac{[5]}{3}$
y	$\frac{1}{2}$	1	0	$-\frac{1}{6}$	$\frac{5}{18}$	0	1.5	$[10] = [6] - \frac{[9]}{2}$
t	8	0	0	1	-1	1	264	$[11] = [7] + [9] \times 3$
P	6	0	0	$\frac{1}{3}$	$\frac{4}{9}$	0	150	$[12] = [8] + [9]$

d $P = 150, x = 0, y = 1.5, z = 48, r = 0, s = 0, t = 264$

e The third constraint (t is not zero).

5 a

	D	E	F	Available
A	20	4	–	24
B	–	26	6	32
C	–	–	14	14
Required	20	30	20	

b Improvement indices $I_{13} = -2, I_{21} = -2, I_{31} = -13,$
$I_{32} = -12$

c Entering cell $(3, 1)$, exiting cell $(3, 3)$. Cost £1384.

	D	E	F	Available
A	6	18	–	24
B	–	12	20	32
C	14	–	–	14
Required	20	30	20	

6 a e.g. a company with several retail outlets supplied from several warehouses, so the cost of transport depends on the source and destination.

b Not balanced (supply = 120, demand = 110)

c Cost 545

	d	e	Dummy	Supply
A	45	–	–	45
B	5	30	–	35
C	–	30	10	40
Demand	50	60	10	120

d Shadow costs $R_1 = 0, R_2 = -1, R_3 = -3, K_1 = 5,$
$K_2 = 7, K_3 = 3$
Improvement indices $I_{12} = -4, I_{13} = -3, I_{23} = -2, I_{31} = 0$

e Entering cell $(1, 2)$, exiting cell $(2, 2)$. Cost 425.

	d	e	Dummy	Supply
A	15	30	–	45
B	35	–	–	35
C	–	30	10	40
Demand	50	60	10	120

7 a Degenerate for an $m \times n$ problem when the allocation occupies fewer than $(m + n - 1)$ cells.

b A dummy location is needed when supply and demand are not equal (unbalanced problem).

c Extra occupied cell $(3, 3)$ created because degenerate.

	1	2	Dummy	Supply
A	15	–	–	15
B	1	11	–	12
C	0	–	7	7
Demand	16	11	7	34

d Shadow costs $R_1 = 0, R_2 = -1, R_3 = 6, K_1 = 62,$
$K_2 = 49, K_3 = -6$
Improvement indices $I_{12} = -2, I_{13} = 6, I_{23} = 7, I_{32} = 3$
Entering cell $(1, 2)$, exiting cell $(2, 2)$. Cost 1497.
(All indices are now positive.)

	1	2	Dummy	Supply
A	4	11	–	15
B	12	–	–	12
C	0	–	7	7
Demand	16	11	7	34

8 x_{ij} is number of thousands of litres allocated to cell (i, j) (all integers)

Minimise $23x_{11} + 31x_{12} + 46x_{13} + 35x_{21} + 38x_{22}$
$+ 51x_{23} + 41x_{31} + 50x_{32} + 63x_{33}$

subject to $x_{11} + x_{12} + x_{13} = 540$
$x_{21} + x_{22} + x_{23} = 789$
$x_{31} + x_{32} + x_{33} = 673$
$x_{11} + x_{21} + x_{31} = 257$
$x_{12} + x_{22} + x_{32} = 348$
$x_{13} + x_{23} + x_{33} = 412$
$x_{ij} \geqslant 0$ for all i and j

D2

9 a i Row/column reduction gives

	I	II	III	IV
C	0	4	0	4
J	1	2	8	0
N	1	0	2	0
S	1	5	10	0

ii Then modify to give

	I	II	III	IV
C	0	4	0	5
J	0	1	7	0
N	1	0	2	1
S	0	4	9	0

Allocation C-III, J-I, N-II, S-IV or C-III, J-IV, N-II, S-I

b

	I	II	III	IV
C	24	2	8	20
J	23	4	0	24
N	21	4	4	22
S	25	3	0	26

10 a Row/column reduction gives

	Cutting	Stitching	Filling	Dressing
A	2	5	13	0
B	0	0	0	0
C	1	3	1	0
D	4	7	7	0

Modify to

	Cutting	Stitching	Filling	Dressing
A	1	4	12	0
B	0	0	0	1
C	0	2	0	0
D	3	6	6	0

and then to

	Cutting	Stitching	Filling	Dressing
A	0	3	11	0
B	0	0	0	2
C	0	2	0	1
D	2	5	5	0

Allocate A cutting, B stitching, C filling, D dressing.
b 270 seconds

11 Subtract all entries from largest.

	H	I	P	S
A	11	6	2	17
B	14	7	0	15
C	11	5	3	15
D	17	9	4	21

Reduce rows/columns:

	H	I	P	S
A	1	2	0	3
B	6	5	0	3
C	0	0	0	0
D	5	3	0	5

Modify to

	H	I	P	S
A	0	1	0	2
B	5	4	0	2
C	0	0	1	0
D	4	2	0	4

and then to e.g.

	H	I	P	S
A	0	1	2	2
B	3	2	0	0
C	0	0	3	0
D	2	0	0	2

Allocate A-H, B-P, C-S, D-I or A-H, B-S, C-I, D-P (income £1077)

12 a The sum of the gains for all players is zero.

b

	I	II	III	Row minima	Max of row minima
I	5	2	3	2	
II	3	5	4	3	3
Column maxima	5	5	4		
Min of column maxima			4		

Max of row minima is not equal to min of row maxima, so no stable solution

c A plays I with probability p. Expected gains are $2p + 3$, $5 - 3p$ and $4 - p$.
Optimal at intersection of $2p + 3$ and $4 - p$, giving $p = \frac{1}{3}$, value $= 3\frac{2}{3}$

d B plays I, II, III with probabilities q_1, q_2, q_3.
$x_1 = \frac{q_1}{v}$, $x_2 = \frac{q_2}{v}$, $x_3 = \frac{q_3}{v}$
Maximise $P = x_1 + x_2 + x_3$ subject to
$5x_1 + 2x_2 + 3x_3 \leqslant 1$, $3x_1 + 5x_2 + 4x_3 \leqslant 1$, $x_1, x_2, x_3 \geqslant 0$

13 a Row minima are $-5, -1, -4, -1$, max $= -1$. Column maxima are $0, 5, -1, 4$, min $= -1$. Play-safe A plays II or IV, B plays III.

b max of row mins = min of col maxs, so there is a stable solution.
Saddle point (II, III) and (IV, III)

c Value of game to B is 1.

14 a Row mins $-1, -4$, max $= -1$. Col maxs $2, 4, 3$, min $= 2$. These are not equal so no stable solution.

b Denis plays I with probability p. Expected gains are $5p - 3$, $4 - 5p$, $7p - 4$.
Optimal when $5p - 4 = 4 - 5p$, so $p = 0.7$.
Value $= 0.5$.

15 A plays I and II with probabilities $\frac{11}{17}$, $\frac{6}{17}$. B plays I and II with probabilities $\frac{8}{17}$, $\frac{9}{17}$.
Value of game $= \frac{14}{17}$

Chapter 5

Exercise 5.1

1 a No **b** Yes, e.g. *ABCGFEDA*
  c Yes, e.g. *ADBECFA* **d** Yes, e.g. *ACBFHGEDA*
2 e.g. *ACBDGFEA*
3 a i 1 **ii** 3 **iii** 12 **b** $\frac{1}{2}(n-1)!$

Exercise 5.2

1 a

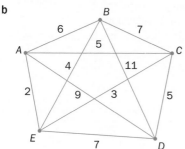

b *ABCDA* = 27 + 20 + 37 + 44 = 128
  ABDCA = 27 + 17 + 37 + 47 = 128
  ACBDA = 47 + 20 + 17 + 44 = 128
c *ABCBDBCA*
2 a *ABCDEA* = 29, *ABCEDA* = 37, *ABECDA* = 30,
  AEBCDA = 32. Min = 29.
b

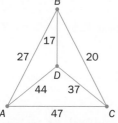

c *AECDBA* = 27
d *AECDEBA* = 27

Exercise 5.3

1 a

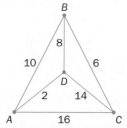

b *ADBCA*, *BCDAB*, *CBDAC*, *DABCD* give 2
  distinct tours *ABCDA*, *ADBCA* of length 32
c *ADBCBDA*
2 a *AEBDCA* 38 **b** *BDAECB* 35
3 a Radstock-Frome-Shepton Mallet-Wells-Cheddar-
  Glastonbury-Radstock 107 km
b Frome-Radstock-Shepton Mallet-Wells-Cheddar-
  Glastonbury-Frome 105 km
  Visit in order Radstock-Shepton Mallet-Wells-
  Cheddar-Glastonbury-Frome-Radstock or vice
  versa.

4 a

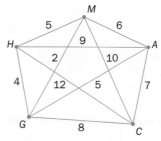

b *HGMACH* = 31 minutes
  On original network *HGMACGH*

5 a

	A	B	C	D
A	–	5	9	6
B	5	–	10	7
C	9	10	–	3
D	6	7	3	–

b *ABDCA*, 24 miles. Actual route *ABDCDA*
6 e.g. *ABCEDA*, £980

Exercise 5.4

1 a *MST* is {*AB*, *AC*, *AD*}, total 42. Upper bound = 84
  minutes.
b e.g. replace *CAB* by *CB* and *BAD* by *BD*. Route is
  ACBDA = 67 minutes.
2 a *MST* is {*AC*, *BC*, *BE*, *CF*, *DE*}, total 34. Upper
  bound = 68 km.
b e.g. replace *ACB* by *AB* and *DEBCF* by *DF*. Route
  is *ABEDFCA* = 50 km
3 a {*AE*, *BC*, *BE*, *DE*} or {*AE*, *BC*, *CE*, *DE*}, both 74.
b 148
c e.g. from second *MST* replace *AED* by *AD* and
  DECB by *DB*.
  Route is *ADBCEA* = 103.
4 a *MST* is {*AC*, *BD*, *CD*, *CF*, *DE*}, total 28.
b 56
c e.g. Replace *DCF* by *DF*, *ACDB* by *AB*, *EDB* by *BE*.
  Upper bound = 41.

5 a

	A	B	C	D	E	F
A	–	11	10	13	10	7
B	11	–	8	15	12	10
C	10	8	–	7	4	9
D	13	15	7	–	6	12
E	10	12	4	6	–	9
F	7	10	9	12	9	–

b {*AF*, *BC*, *CE*, *CF*, *DE*} or {*AF*, *BC*, *CE*, *DE*, *EF*},
  both 34 miles
c 68 miles
d e.g. replace *BCEFA* by *BA* and *FED* by *FD*. Route
  is *AFDECBA* = 48 miles
e *AGFGHDEHCBGA*, 3 times at *G*, twice at *H*.

Exercise 5.5

1 **a** Edges $(AC + AD) + (BC + CD) = 14$. Not possible because a tour of length 14 would have to use AC, CD and AD, which form a cycle.

b Edges $(BC + BD) + (AC + AD) = 16$.
These form a cycle $ACBDA$, which must therefore be an optimal tour.

2 Bounds obtained are 63, 59, 72, 60, 55, 57. Best is 72, found from deleting C.

3 **a**

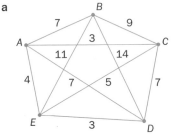

b Alphabetical tour length 3520 m
Lower bound deleting A is 3400 m.

4 Lower bounds are 74, 72, 72, 77, 72. Best is 77 by deleting D.

5 **a** Best lower bound \$400 (deleting C)

b $\$400 \leqslant T \leqslant \$420$, where T is the cost of the optimal tour.

6 Lower bounds are 169, 169, 169, 193, 209. Best is 209 km by deleting Stoke on Trent.

7 **a**

	A	B	C	D	E
A	–	3	6	9	4
B	3	–	6	7	5
C	9	6	–	9	7
D	6	7	9	–	2
E	4	5	7	2	–

b Best nearest neighbour tour = 24
Best lower bound = 22.

8 **a** Best upper bound = best lower bound = 44 points.
Optimal route $SBACEDS$

b $81 \leqslant$ best route $\leqslant 84$. $SCBDAES = 81$ points.

9 **a**

	A	B	C	D	E	F
A	–	4	2	4	3	5
B	4	–	2	4	3	5
C	2	2	–	4	3	3
D	4	4	4	–	3	3
E	3	3	3	3	–	2
F	5	5	3	3	2	–

b Upper bound 16, lower bound 15.

Review 5

1 **a**

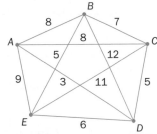

b $ADCBEA = 29$ **c** $ADCBEDA$

2 **a** $ACDEBA = 37$ **b** 30

c 33, which is better than part **b**

d $33 \leqslant T \leqslant 37$

3 **a** MST is $\{AE, BC, BD, BE, BG, EF\} = 29$ km. Upper bound = 58 km.

b e.g. Replace $DBEF$ by DF and AEF by AF. Route $AFDBGCBEA = 49$ km.

c Lower bound = 34. $34 \leqslant T \leqslant 49$

4 **a** MST is $\{PQ, QS, RT, ST, SU\} = 35$ mins. Upper bound = 70 mins.
Short cuts e.g. Replace $PQSU$ by PU and $RTSU$ by RU gives 49 mins.

b 46 mins.

5 **a**

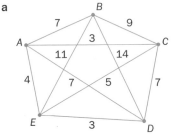

b $ACEDBA = 32$ miles.
Visit in order $ACEDEABA$

c Best = 31 miles   **d** $ABCDEA = 30$ miles

e 24 miles, 26 miles   **f** $26 \leqslant T \leqslant 30$

Chapter 6

Exercise 6.1

1 No edge flow > capacity. Inflow = outflow at $A(12)$, $B(17)$, $C(17)$ and $D(19)$.

2 **a** 7   **b** 23   **c** 2, 4, 6

d No branching, effectively a single arc of capacity 10.

3 **a** i 8, 6   ii 9   **b** 6

4 **a** Outflow at S ($w + 30$) must equal inflow at T (50)

b 5, 4, 31   **c** AD, ET

Exercise 6.2

1 **a** i $\{SA, SB, SC\}$, $\{SC, SB, AB, AT\}$, $\{SA, BT, CT, (AB)\}$

ii $X = \{S\}$, $Y = \{A, B, C, T\}$: $X = \{S, A\}$, $Y = \{B, C, T\}$: $X = \{S, B, C\}$, $Y = \{A, T\}$

iii 41, 52, 38

b i $\{SB, AB, CF, EF, ET, (BC, DC)\}$, $\{SA, BC, BD, (AB)\}$, $\{AC, BC, DC, FT, (CF, EF)\}$

ii $X = \{S, A, C, E\}$, $Y = \{B, D, F, T\}$: $X = \{S, B\}$, $Y = \{A, C, D, E, F, T\}$: $X = \{S, A, B, D, F\}$, $Y = \{C, E, T\}$

iii 90, 70, 69

c i $\{SA, SB, CE, (AC, BC)\}$, $\{BF, CE, DE, DT\}$, $\{AD, ET, FT, (DE)\}$

ii $X = \{S, C\}$, $Y = \{A, B, D, E, F, T\}$: $X = \{S, A, B, C, D\}$, $Y = \{E, F, T\}$: $X = \{S, A, B, C, E, F\}$, $Y = \{D, T\}$

iii 144, 106, 107

d i $\{SB, SC, AC, AD\}$, $\{SC, BE, AC, DC, DT, (BC)\}$, $\{AD, CT, ET, (DC)\}$

ii $X = \{S, A\}$, $Y = \{B, C, D, E, T\}$: $X = \{S, A, B, D\}$, $Y = \{C, E, T\}$: $X = \{S, A, B, C, E\}$, $Y = \{D, T\}$:

iii 104, 131, 118

2 a i $\{SA, SB, CT\}$, $X = \{S, C\}$, $Y = \{A, B, T\}$
ii SAT 12, SBT 9, SCT 14

b The cut in **a i** is minimum, the flow in **a ii** is maximum.

3 $\{CE, FT\}$, $X = S, A, B, C, D, F$, $Y = \{E, T\}$, capacity 24. Flow $SACET$ (10) + $SBDFT$ (14) = 24.

4 Cut $\{SA, BC, BE, (AB)\}$, $X = \{S, B\}$, $Y = \{A, C, D, E, T\}$, capacity 17.
Flow $SACT$ (6) + $SBET$ (5) + $SBCET$ (3) + $SADT$ (3) = 17 (other flow patterns possible).

Exercise 6.3

1 a Flow-augmenting paths $SABT$ (2), $SABCT$ (4). Total flow 45

b $\{AT, AB, SB, SC\} = 45$

2 a Flow-augmenting paths $SADET$ (10), $SACET$ (10). Total flow 100

b $\{AD, CE, BF\} = 100$

3 Flow-augmenting path $SBCADET$ (10)

4 a Flow-augmenting paths $SACT$ (2), $SBCET$ (2), $SABCET$ (1), $SADCT$ (1).

b e.g. initial flow $SADT$ (24), SCT (38), $SBET$ (40). Flow-augmenting paths $SCET$ (14), $SACBET$ (2). Flow = 118. Cut $\{AD, CT, ET, (CD)\} = 118$

Exercise 6.4

1 a Sources A, E **b** Additional arcs SA (9), SE (14)

c e.g. initial flow $SEFG$ (8), $SACG$ (3), $SADG$ (2). Flow-augmenting path $SEBFGDT$ (2). Flow = 15.

d

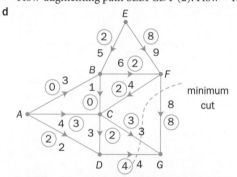

2 a Additional arcs SA (8), SB (16), HT (14), IT (9)

b 5, 5, 6 **c** Flow-augmenting path $SADGHT$ (3)

d e.g. $\{EH, DG, (FG), FI\} = 19$

3 a Additional arcs SB (36), SD (26), IT (14), KT (19)

b e.g. initial flow $SBAEIT$ (6), $SDHLKT$ (10), $SDCGKT$ (5), $SBFJIT$ (5). Flow-augmenting path $SBFJKT$ (2). Value of flow = 28.

c e.g. $\{AE, (EF), FJ, GK, KL\} = 28$

Review 6

1 a e.g. cut $\{AB, AD, CD, CE\} = 25$

b i 6 **ii** 10

c

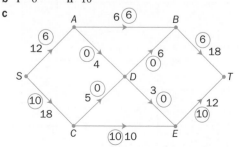

d e.g. Flow-augmenting paths $SADBT$ (4), $SCDBT$ (2), $SCDET$ (2). Max flow 24.

e e.g.

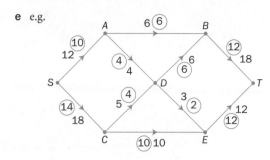

f $\{AB, BD, ET\} = 24$

2 a Sources A, E Sinks D, H

b

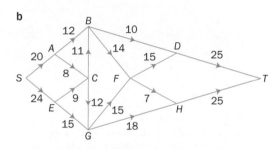

c e.g. initial flow $SABDT$ (10), $SACBFDT$ (8), $SEGHT$ (15), $SECGFHT$ (7)
Flow-augmenting paths $SABFDT$ (2), $SECGHT$ (2). Max flow = 44
Saturated edges SA, SE, AB, AC, EC, EG, BD, FH. Maximal because e.g. $\{AB, AC, EC, EG\} = 44$ is a minimum cut.

3 a A, E, G **b** 45

c e.g. EHD (2), $ECHD$ (1)

d

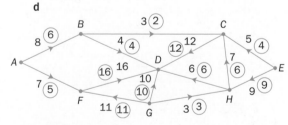

Max flow = 48

e Cut $\{BD, CD, HD, GD, FD\} = 48$

4 a 35, 26, 31

b Because there is a cut and a flow both of value 26, the max flow–min cut theorem states that this flow is maximal.

c EJ (5) increases total flow by 1. FH (3) increases total flow by 2. Option 2 is the better one.

5 a $x = 3$, $y = 26$

b SF_1 (42), SF_2 (41), R_1T (47), R_2T (32), R_3T (10)

c e.g. SF_1AER_1T (7), SF_1BER_1T (5), SF_1BGR_1T (1), SF_2CDBGR_2T (4)

d

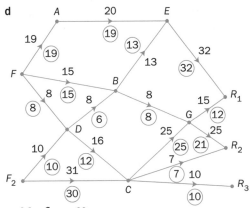

Max flow = 82

e e.g. $\{F_1A, BE, BG, CG, CR_2, CR_3\} = 82$

6 a

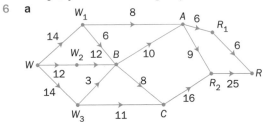

b i 6 **ii** 11

c e.g. WW_1BAR_2R (6), WW_1AR_2R (2), WW_2BCR_2R (5), WW_2BAR_2R (1). Max flow = 31 **d** 12

Chapter 7
Exercise 7.1

1 $SBDHT$ (61) **2** $SADGT$ (82)

3 a, b

[figure]

c PRQ, £61 000

4 a

[figure]

b $SAFT$ or $SCET$, £20

D2

Exercise 7.2

1

Stage	State	Action	Destination	Value
1	G	G-T	T	18*
	H	H-T	T	22*
	I	I-T	T	15*
2	D	D-G	G	27*
		D-H	H	34
	E	E-H	H	35
		E-I	I	23*
	F	F-G	G	25
		F-I	I	23*
3	A	A-D	D	41
		A-E	E	39
		A-F	F	33*
	B	B-D	D	42
		B-E	E	41*
	C	C-E	E	40
		C-F	F	37*
4	S	S-A	A	40*
		S-B	B	51
		S-C	C	43

Shortest route $SAFIT$ (40)

2

Stage	State	Action	Destination	Value
1		E-T	T	29*
		F-T	T	26*
		G-T	T	19*
2		A-E	E	45
		A-F	F	49*
		A-G	G	33
		B-E	E	49*
		B-F	F	47
		B-G	G	45
		C-E	E	51*
		C-F	F	41
		C-G	G	47
		D-E	E	51*
		D-F	F	44
		D-G	G	39
3		S=A	A	76
		S-B	B	74
		S-C	C	79*
		S-D	D	75

Max weight $SCET$ 79 tonnes

3

Stage	State	Action	Destination	Value
1	AB	C	ABC	230*
	AC	B	ABC	270*
	BC	A	ABC	280*
2	A	B	AB	510*
		C	AC	540
	B	A	AB	470*
		C	BC	550
	C	A	AC	510*
		B	BC	560
3	None	A	A	720
		B	B	660*
		C	C	710

Best order BAC 660 euros

4

Stage	State (stock)	Action (make)	Destination (stock)	Value
1	0	2	0	450*
	1	1	0	250*
	2	0	0	0*
2	0	3	0	1100*
		4	1	1130
	1	2	0	900*
		3	1	930
		4	2	910
	2	1	0	700*
		2	1	730
		3	2	710
3	1	4	0	1950*
	2	3	0	1750*
		4	1	1780
4	0	3	1	2630*
		4	2	2660

Make week 1–3 sets, week 2–4 sets, week 3–3 sets, week 4–2 sets.

Exercise 7.3

1 a

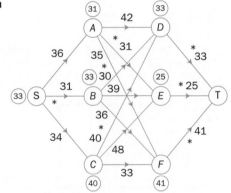

Minimax route *SBDT*. Longest leg 33 km

b Yes, 94 km.

2

Stage	State	Action	Destination	Value
1	D	D-T	T	33*
	E	E-T	T	25*
	F	F-T	T	41*
2	A	A-D	D	33
		A-E	E	25
		A-F	F	35*
	B	B-D	D	30
		B-E	E	25
		B-F	F	36*
	C	C-D	D	33*
		C-E	E	25
		C-F	F	33*
3	S	S-A	A	35*
		S-B	B	31
		S-C	C	33

Maximin route *SAFT*. Min value 35.

3

Stage	State	Action	Destination	Value
1	G	G-T	T	16*
	H	H-T	T	12*
	I	I-T	T	12*
2	D	D-G	G	16*
		D-H	H	12
	E	E-H	H	12*
		E-I	I	12*
	F	F-G	G	16*
		F-I	I	12
3	A	A-D	D	16*
		A-E	E	12
		A-F	F	16*
	B	B-D	D	16*
		B-F	F	16*
	C	C-E	D	16*
		C-E	E	12
		C=F	F	16*
4	S	S-A	A	15
		S-B	B	10
		S-C	C	16*

Heaviest 16 tonnes. Route *SCDGT* or *SCFGT*

4 a

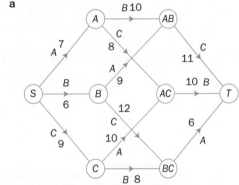

b

Stage	State	Action	Destination	Value
1	AB	C	T	11*
	AC	B	T	10*
	BC	A	T	6*
2	A	B	AB	10*
		C	AC	8
	B	A	AB	9*
		C	BC	6
	C	A	AC	10*
		B	BC	6
3	S	A	A	7
		B	B	6
		C	C	9*

Optimal order *CAB*. Min profit £9000

D2

Review 7

1 a, b

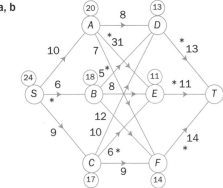

Route *SBDT*, distance 24 miles

2 a

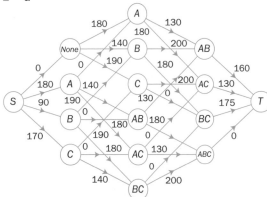

b

Stage	State	Action	Destination	Value
1	AB	C	T	160*
	AC	B	T	130*
	BC	A	T	175*
	ABC	None	T	0*
2	A	B	AB	290*
		C	AC	310
	B	A	AB	360
		C	BC	355*
	C	A	AC	330
		B	BC	305*
	AB	C	ABC	180
		None	AB	160*
	AC	B	ABC	130*
		None	AC	130*
	BC	A	ABC	200
		None	BC	175*
3	A	B	AB	300
		C	AC	320
		None	A	290*
	B	A	AB	340*
		C	BC	365
		None	B	355
	C	A	AC	310
		B	BC	315
		None	C	305*
	None	A	A	470*
		B	B	495
		C	C	495
4	S	A	A	470
		B	B	430*
		C	C	475
		None	None	470

Optimal schedule *B*, *A*, None, *C*. Cost £430

3 a

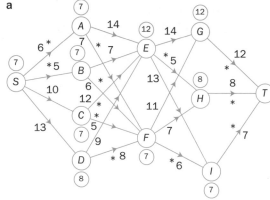

Minimax route *SAFIT* or *SBFIT*, value 7.

b

Stage	State	Action	Destination	Value
1	G	G-T	T	12*
	H	H-T	T	8*
	I	I-T	T	7*
2	E	E-G	G	12*
		E-H	H	
		E-I	I	7
	F	F-G	G	11*
		F-H	H	7
		F-I	I	6
3	A	A-E	E	12*
		A-F	F	7
	B	B-E	E	7*
		B-F	F	6
	C	C-E	E	12*
		C-F	F	5
	D	D-E	E	9*
		D-F	F	8
4	S	S-A	A	6
		S-B	B	5
		S-C	C	10
		S-D	D	9

Maximin route *SCEGT*, value 10

4

Stage	State (stock)	Action (make)	Destination (stock)	Value
1	1	6	0	960*
	2	5	0	810*
2	0	6	1	1940*
	1	5	1	1790*
		6	2	1810
	2	4	1	1640*
		5	2	1660
3	0	2	0	2300*
		3	1	2320
		4	2	2340
	1	1	0	2150*
		2	1	2170
		3	2	2190
	2	0	0	1940*
		1	1	2020
		2	2	2040
4	0	4	0	2960
		5	1	2980
		6	2	2940*

Optimal schedule: week 1–6, week 2–0, week 3–6, week 4–6. Cost £2940.

5 a, b

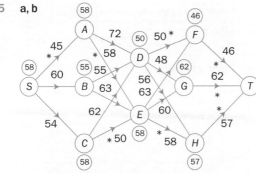

Minimax route *SAEHT*, largest amount of water 58 litres.

6

Stage	State	Action	Destination	Value
	AB	C	T	30*
1	AC	B	T	28*
	BC	A	T	25*
	A	B	AB	54
		C	AC	56*
2	B	A	AB	53*
		C	BC	51
	C	A	AC	52*
		B	BC	51
		A	A	72
3	S	B	B	75
		C	C	77*

Optimal order *C, A, B*. Profit £77 000.

Revision 2

1 a

	A	B	C	D	E	F
A	0	20	30	32	12	15
B	20	0	10	25	32	16
C	30	10	0	15	35	19
D	32	25	15	0	20	34
E	12	32	35	20	0	16
F	15	16	19	34	16	0

b $AEFBCDA = 12 + 16 + 16 + 10 + 15 + 32 = 101$ upper bound.

c *AEFBCDEA*

d e.g. *BCDEAFB* = 88

2 a

	1	7	6	5	2	4	3
	A	B	C	D	E	F	G
A	–	165	195	280	130	200	150
B	165	–	⑨⓪	155	150	235	230
C	195	90	–	170	⑪⑩	175	190
D	280	155	170	–	150	⑩⑤	163
E	⑬⓪	150	110	150	–	90	82
F	200	235	175	105	90	–	⑥③
G	150	230	190	163	⑧②	63	–

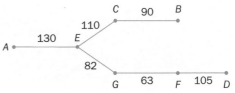

Order in which arcs are selected: *AE, EG, GF, FD, EC, CB*

b i Initial upper bound $= 2 \times$ (weight of *MST*)
$= 2 \times 580 = 1160\,\text{km}$

ii Use *BD* instead of *BCEGFD*.
Saving $(90 + 110 + 82 + 63 + 105) - 155 = 295$
New upper bound $= 865\,\text{km}$. Route is *AECBDFGEA*.

iii Use *AG* instead of *AEG*. Saving $(130 + 82) - 150 = 62\,\text{km}$. Length $= 803\,\text{km}$. Route is *AECBDFGA* which visits each vertex once only.

3 a The table corresponds to a network, with arc weights giving the changeover times. A tour corresponds to a particular order of activities, and the order with the least total weight is required.

b $480\,\text{min} = 8\,\text{h}$

c e.g. *BFTCDB* = 396 min = 6 h 36 min

d Shortest arcs from *B* are 64 and 100. Minimum connector of remaining graph is *CD, CT, FT* = 54 + 60 + 68 = 182. Lower bound = 346 min

4 a

b i $SF_1ABR = 6$, $SF_3CR = 8$ **ii** See diagram

c i e.g. $SF_1BR = 6$, $SF_2BR = 3$, $SF_2CR = 3$, $SF_3R = 4$
Total flow = 30

ii Cut $\{BR, F_2C, F_3C, F_3R\} = 30$ so max flow–min cut theorem applies.

5 a $x = 9$, $y = 16$

b Initial flow 53. Not maximal because e.g. there is a flow-augmenting path *IDA*.

c e.g. $IDA = 9$, $IFDA = 2$. Total flow = 64.

d

e The cut $\{GC, FA, (DF), ID, IE, HE\} = 64$, so max flow–min cut theorem applies.

6 a

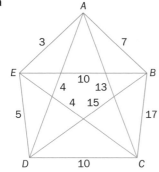

b $AEDBCA = 45$ km **c** $AEDBDCA$

7 a $C_1 = 83$, $C_2 = 177$

b, c

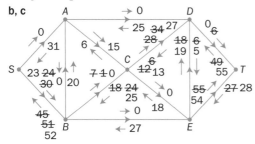

Flow-augmenting paths $SBCDT$ (6), $SBCDET$ (1)
as shown – other versions are possible.
Total flow = 83.

d

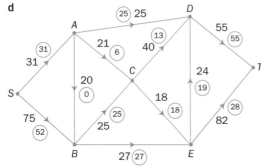

e Flow = capacity of cut C_1, so max flow–min cut
theorem applies.

8 a Total cost = $2 \times 40 + 350 + 200 = £630$

b

Stage	State	Action	Destination	Value
1	0	2	0	200*
	1	1	0	240*
2	1	4	0	590 + 200 = 790*
	2	3	0	280 + 200 = 480*
		4	1	630 + 240 = 870
3	0	4	1	550 + 790 = 1340*
	1	3	1	240 + 790 = 1030*
		4	2	590 + 480 = 1070
4	0	3	0	200 + 1340 = 1540*
		4	1	550 + 1030 = 1580

(This table adds the cost of storage during the
month after the cycles were made.)

Month	August	September	October	November
Make	3	4	4	2

Cost £1540.

c Profit = £14 660

9

Stage	Initial state	Action	Final state	Value
1	D	DT	T	30
	E	ET	T	40
	F	FT	T	20
2	A	AD	D	max (40, 30) = 40*
		AE	E	max (55, 40) = 55
	B	BD	D	max (50, 30) = 50*
		BE	E	max (60, 40) = 60
	C	CD	D	max (45, 30) = 45
		CE	E	max (50, 40) = 50
		CF	F	max (30, 20) = 30*
3	S	SA	A	max (40, 20) = 40*
		SB	B	max (50, 30) = 50
		SC	C	max (40, 30) = 40*

There are two routes: $SCFT$ and $SADT$
Max altitude = 40 (× 100 ft) = 4000 ft.

10 a Stage – number of weeks to finish, state – current
show, action – next show

b

Stage	State	Action	Destination	Value
1	F	F–Home	Home	500 – 80 = 420 *
	G	G–Home	Home	700 – 90 = 610 *
	H	H–Home	Home	600 – 70 = 530 *
2	D	DF	F	1500 – 200 + 420 = 1720
		DG	G	1500 – 160 + 610 = 1950 *
		DH	H	1500 – 120 + 530 = 1910
	E	EF	F	1300 – 170 + 420 = 1550
		EG	G	1300 – 100 + 610 = 1810 *
		EH	H	1300 – 110 + 530 = 1720
3	A	AD	D	900 – 180 + 1950 = 2670 *
		AE	E	900 – 150 + 1810 = 2560
	B	BD	D	800 – 140 + 1950 = 2610 *
		BE	E	800 – 120 + 1810 = 2490
	C	CD	D	1000 – 200 + 1950 = 2750 *
		CE	E	1000 – 210 + 1810 = 2600
4	Home	Home–A	A	–70 + 2670 = 2600*
		Home–B	B	–80 + 2610 = 2530
		Home–C	C	–150 + 2750 = 2600*

c She should travel from home to A or C, then
to D, then to G and finally home, making £2600
profit.

Glossary

Key words are given in bold type. The black text in this glossary is taken from the Edexcel glossary. Additional explanation is given in blue type.

1 Further linear programming

The **simplex tableau** for the linear programming problem:

Maximise $\quad P = 14x + 12y + 13z$,

Subject to $\quad 4x + 5y + 3z \leqslant 16$,

$\qquad\qquad\ 5x + 4y + 6z \leqslant 24$,

will be written as

Basic variable	x	y	z	r	s	Value
r	4	5	3	1	0	16
s	5	4	6	0	1	24
P	-14	-12	-13	0	0	0

where r and s are slack variables.

The **basic variables** are non-zero, the other (non-basic) variables are zero.

The **optimality condition** – the solution is optimal if there are no negative entries in the objective row.

If the solution is not optimal, you **change the basis**, using row operations on the tableau. This uses the **simplex algorithm** (see Chapter 1 for details).

2 Transportation problems

Each **source** has a number of units of stock – its **supply**, **capacity** or **availability**.

Each **destination** needs a number of units – its **demand** or **requirement**.

In the **north-west corner method**, the upper left-hand cell is considered first and as many units as possible sent by this route.

You continue by moving down and across the tableau, at each stage allocating as many units as are needed or available.

The **shadow costs** R_i, for the ith row, and K_j, for the jth column, are obtained by solving $R_i + K_j = C_{ij}$ for **occupied cells**, taking $R_1 = 0$ arbitrarily.

The **improvement index** I_{ij} for an **unoccupied cell** is defined by $I_{ij} = C_{ij} - R_i - K_j$. An optimal allocation can be made if there are no negative improvement indices.

The **stepping stone method** is an iterative procedure for moving from an initial feasible solution to an optimal solution. This allocates units to a cell with a negative improvement index by means of a sequence of adjustments round a loop.

A problem is **balanced** if total supply equals total demand. If the problem is unbalanced you add a dummy row or column, with zero unit costs, to balance the problem.

Degeneracy occurs in a transportation problem, with m rows and n columns, when the number of occupied cells is less than $(m + n - 1)$. You create extra occupied cells by inserting zeros.

3 Allocation (assignment) problems

A **cost matrix** shows the cost of possible pairing of two sets. A **payoff matrix** shows the gains from possible pairings. You convert a payoff matrix to a cost matrix by subtracting all entries from the largest entry.

An **opportunity cost matrix** is produced from a cost matrix by **row reduction** (subtract the minimum value in each row from every cell in the row) and then **column reduction** (subtract the minimum value in each column from every cell in the column).

An optimal assignment can be made in an $n \times n$ opportunity cost matrix if you need n straight lines to cover all the zeros. If no assignment possible, draw lines to cover zeros and modify table by subtracting the least uncovered number from all uncovered numbers and adding it to cells at intersection of lines.

The process of successively modifying a table and testing for optimality is the **Hungarian algorithm**.

4

A **two-person game** is one in which only two parties can play – that is, a game with two 'sides'.

A **zero-sum** game is one in which the sum of the losses for one player is equal to the sum of the gains for the other player.

The **payoff matrix** shows the gains for the player on the left of the table – the row player.

A **play safe** strategy gives the best guaranteed outcome.

The **value** of a game is the payoff to the row player if both players use their best strategy.

A game has a **stable solution** (a **saddle point**) if neither player can gain by changing from their play safe strategy. In this case **max of row minima = min of column maxima = value** of the game.

A row or column can be eliminated if it is **dominated** by another.

row i dominates row j if, for every column, value in row $i \geqslant$ value in row j

column i dominates column j if, for every row, value in col $i \leqslant$ value in col j

5 The travelling salesman problem

The **travelling salesman problem** is 'find a route of minimum length which visits every vertex in an undirected network'. In the '**classical**' problem, each vertex is visited once only. In the '**practical**' problem, a vertex may be revisited.

For three vertices A, B and C, the **triangular inequality** is
'length $AB \leqslant$ length $AC +$ length CB'.

A **walk** in a network is a finite sequence of edges such that the end vertex of one edge is the start vertex of the next.

A walk which visits every vertex, returning to its starting vertex, is called a **tour**.

A **Hamiltonian cycle (tour)** is a closed path which visits every vertex of the graph. A **Hamiltonian graph** has at least one Hamiltonian cycle.

In the **nearest neighbour algorithm** you travel at each stage to the nearest unvisited vertex.

6 Flows in networks

A **cut**, in a network with source S and sink T, is a set of arcs (edges) whose removal separates the network into two parts X (the **source set**) and Y (the **sink set**), where X contains at least S and Y contains at least T. The **capacity of a cut** is the sum of the capacities of those arcs in the cut which are directed from X to Y.

The **maximum flow – minimum cut theorem** states:

The value of the maximal flow = the capacity of a minimum cut

A **flow-augmenting path** is a route from S to T where all edges have spare capacity.

If a network has several sources $S_1, S_2, \ldots$, then these can be connected to a single **supersource** S. The capacity of the edge joining S to S_1 is the sum of the capacities of the edges leaving S_1.

If a network has several sinks $T_1, T_2, \ldots$, then these can be connected to a **supersink** T. The capacity of the edge joining T_1 to T is the sum of the capacities of the edges entering T_1.

7 Dynamic programming

Bellman's principle for dynamic programming is 'Any part of an optimal path is optimal.'

The **minimax route** is the one in which the maximum length of the arcs used is as small as possible.

The **maximin route** is the one in which the minimum length of the arcs used is as large as possible.

Index

actions 141
allocation problems 53-68
   linear programming
     formulation 64
   maximisation problems 60-1
   opportunity cost matrix 54-5
   testing and revising 56-8
   unbalanced problems 62
assignment problems 53-68
augmentation 128
availability 22

balanced problem 22, 23, 48
basic feasible solution 7
basic non-feasible solutions 7
basic solution 7
basic variables 7
basis
   changing 8
   leaving 8
Bellman's principle of optimality 141
bipartite graph 22
bottleneck 124

capacitated network 120
capacity 22
   of a cut 124
   of the edge 120
cells
   entering 34, 38, 39
   exiting 34, 39
   occupied 28, 29, 30
changing the basis 8
circuit 98
closed path 34
column
   dummy 60, 62
   reduction 54, 57, 60
conservative condition 120, 122
constraints 2, 48
   independent 64
cost matrix 54, 60
cut set 125
cuts 124-6
cycle 98

decision variables 2, 64
degeneracy 28, 42-5
demand 22

destinations 22, 38
Dijkstra's algorithm 48, 98, 100, 140
distribution plan 23-4
dominance 74
dummy column 60, 62
dummy source 38
dynamic programming 139-56
   multi-stage problems 140-2
   using a table 144-7

entering cell 34, 38, 39
entering route 34
entering the basis 8
exiting cell 34, 39
exiting route 34
expectation 76
expected mean 76
expected payoff 76

feasibility condition 120
feasible region 4
final answer 82
flow
   in the edge 120
   in the network 120
flow-augmentation 128
flow-augmenting path 128, 129, 130

game theory 69-88
   dominance 74
   graphical methods for $2 \times n$ and
    $n \times 2$ games 78-9
   linear programming
    formulation 82-4
   mixed strategy games 76-7
   play safe strategy and stable
    solution 70-2
game value 76
graph
   $2 \times n$ and $n \times 2$ games 78-9
   bipartite 22
   linear programming solutions 4
greedy algorithm 102

Hamiltonian cycle (tour)
   98, 100, 102, 110
Hamiltonian graph 98
heuristic algorithm 102
Hungarian algorithm 54, 60

improvement index (indices)
   28, 29, 31, 35-6, 38-40,
   43, 44, 45
independent constraints 64

Kruskal's algorithm 106

labelling procedure 129
leaving the basis 8
linear programming 1-20, 77
   algebraic solutions 6-13
   formulation 2
   game theory 82-4
   graphical solutions 4
   in transportation problem 48
   opportunity cost matrix 64
loop 34, 35
lower bound 106, 110-12

matrix
   cost 54, 60
   opportunity cost 54-5
   payoff 60, 70, 74
maximal flow 128, 130
maximin 71, 150-2
maximisation problems 60-1
maximum flow—minimum cut
   theorem 125, 126, 131
mean payoff 76
minimax 71, 150-2
minimum spanning tree (MST) 106,
   107, 112
mixed strategy game 71, 76
multiple sources and sinks 132

nearest neighbour
   algorithm 102-4
network flows 119-38
non-basic variables 7
north-west corner method 24-5

objective function 2
   maximising 13
   minimising 13
objective line 4
objective row 9
occupied cells 28, 29, 30
opportunity cost matrix 54-5
optimality 28-31

D2